Y0-BYK-902

THIN-LAYER CHROMATOGRAPHY

SCIENTIFIC REPORTS OF THE
ISTITUTO SUPERIORE DI SANITÀ

THIN-LAYER CHROMATOGRAPHY

*Proceedings of the Symposium held at the
Istituto Superiore di Sanità, Rome, 2-3 May, 1963*

EDITED BY

G. B. MARINI-BETTÒLO

Istituto Superiore di Sanità, Rome.

ELSEVIER PUBLISHING COMPANY

AMSTERDAM/LONDON/NEW YORK

1964

ELSEVIER PUBLISHING COMPANY
335 JAN VAN GALENSTRAAT, P.O. BOX 211, AMSTERDAM

AMERICAN ELSEVIER PUBLISHING COMPANY, INC.
52 VANDERBILT AVENUE, NEW YORK, N.Y. 10017

ELSEVIER PUBLISHING COMPANY LIMITED
12B, RIPPLESIDE COMMERCIAL ESTATE
RIPPLE ROAD, BARKING, ESSEX

LIBRARY OF CONGRESS CATALOG CARD NUMBER 64-14182

WITH 105 ILLUSTRATIONS, 40 TABLES AND 557 REFERENCES

ALL RIGHTS RESERVED
THIS BOOK OR ANY PART THEROF MAY NOT BE REPRODUCED IN ANY FORM
INCLUDING PHOTOSTATIC OR MICROFILM FORM,
WITHOUT WRITTEN PERMISSION FROM THE PUBLISHERS

PREFACE

The Istituto Superiore di Sanità decided to organise this symposium on thin-layer chromatography because past experience has shown how useful extended discussion of new and rapidly developing research techniques can be.

Thin-layer chromatography was hardly known and only infrequently employed ten years ago. It owes its present popularity to recent improvements of the technique and though several monographs have appeared of late, it was felt that a meeting of the main research groups would contribute further to its development, as well as permitting a large number of younger research workers to profit from discussions with the experts.

The present volume consists of a series of plenary lectures covering all the principal aspects of thin-layer chromatography and a number of original papers both from the Istituto Superiore di Sanità and from other sources which were represented at the symposium. They serve to illustrate yet again the variety of possible applications of this technique.

In publishing this symposium as a separate volume, the Istituto Superiore di Sanità has departed from its previous practice of publishing such work in the "Scientific Reports of the Istituto Superiore di Sanità" which has now been discontinued. Since the symposia range from pure chemistry to specialised medical topics it was felt that a greater service to the reader would be performed by publishing them separately. This volume is thus the first of a series which will replace the "Scientific Reports". The various volumes will appear as soon as possible after each symposium and will therefore not be bound to particular publication dates, as would a journal.

G. B. Marini-Bettòlo

The present Volume constitutes the Proceedings of the
SYMPOSIUM ON THIN-LAYER CHROMATOGRAPHY

Rome, 2–3 May, 1963

organised

by

Istituto Superiore di Sanità

with the participation of:

M. BARBIER, Gif-sur-Yvette
R. BENIGNI, Milan
H. R. BOLLIGER, Basle
C. BONINO, Bologna
G. BOTTURA, Bologna
A. BRECCIA, Bologna
V. CARELLI, Camerino
C. G. CASINOVI, Rome
G. CAVINA, Rome
M. A. CIASCA, Rome
J. W. COPIUS PEEREBOOM, Leiden
M. COVELLO, Naples
G. CURRI, Padova
G. DE ANGELIS, Rome
E. DEMOLE, Geneva
G. DI MODICA, Torino
P. FASELLA, Rome
M. FÉTIZON, Gif-sur-Yvette
G. GIACOMELLO, Rome
A. GIARTOSIO, Rome
G. GRANDOLINI, Perugia
G. GRASSINI, Rome
M. HRANISAVLJEVIĆ-JAKOVLJEVIĆ, Beograd
L. LÁBLER, Prague
M. LEDERER, Rome
I. LEONCIO D'ALBUQUERQUE, Recife

A. LIBERTI, Naples
G. B. MARINI-BETTÒLO, Rome
K. H. MICHEL, Stockholm
E. G. MONTALVO, Quito
G. M. NANO, Torino
R. NEHER, Basle
A. NIEDERWIESER, Basle
J. OPIENSKA-BLAUTH, Lublin
F. B. PADLEY, St. Andrews
R. PAOLETTI, Milan
G. PATAKI, Basle
I. PEJKOVIĆ-TADIĆ, Beograd
H. J. PETROWITZ, Berlin
E. RAGAZZI, Padova
A. ROMEO, Rome
C. R. ROSSI, Padova
A. ROSSI-FANELLI, Rome
L. SAENZ-PADILLA, Guatemala
F. SANDBERG, Stockholm
O. SCHETTINO, Naples
H. SEILER, Basle
N. SILIPRANDI, Padova
E. STAHL, Saarbrücken
C. TURANO, Rome
C. VICARI, Rome
P. WOLLENWEBER, Dueren

CONTENTS

Preface V

Introductory speech
    by G. Giacomello (Rome) IX

Development and application of thin-layer chromatography
    by E. Stahl (Saarbrücken) 1

Dünnschichtchromatographie auf Cellulose-Schichten
    von P. Wollenweber (Düren) 14

Thin-layer chromatography on loose layers of alumina
    by L. Lábler (Praha) 32

La chromatographie sur couches minces dans le domaine des substances odorantes naturelles et synthétiques
    par E. Demole (Genève) 45

Thin-layer electrophoresis
    by G. Grassini (Rome) 55

Spectrométrie de masse et chromatographie en couche mince
    par M. Fétizon (Gif-sur-Yvette) 69

Thin-layer chromatography of steroids
    by R. Neher (Basel) 75

Thin-layer chromatography of lipids
    by F. B. Padley (St. Andrews) 87

Standardized chromatographic data: A suggestion
    by M. Brenner, G. Pataki and A. Niederwieser (Basel) 116

Chromatography on thin layer of starch with reversed phases
    by J. Davídek (Praha) 117

Centrifugal chromatography. XII. Centrifugal thin-layer chromatography
    by J. Rosmus, M. Pavlíček and Z. Deyl (Praha) 119

Über die Anwendung der Zirkulartechnik beim chromatographieren auf Kieselgel-Dünnschichten. Trennung und Reindarstellung von Morphin, Papaverin und Chinin aus deren Gemischen
    von M. von Schantz (Helsingfors) 122

Zur Dünnschichtchromatographie mehrkerniger aromatischer Kohlenwasserstoffe
    von H.-J. Petrowitz (Berlin-Dahlem) 132

Thin-layer chromatography of 2,4-dinitrophenylhydrazones of aliphatic carbonyl compounds and their quantitative determination
    by G. M. Nano (Torino) 138

Thin-layer chromatography of steroidal bases and Holarrhena alkaloids
    by L. Lábler and V. Černý (Praha) 144

Thin-layer chromatography of alkaloids on magnesia chromatoplates
    by E. Ragazzi, G. Veronese and C. Giacobazzi (Padua) 149

Some applications of thin-layer chromatography for the separation of alkaloids
    by G. Grandolini, C. Galeffi, E. Montalvo, C. G. Casinovi and G. B. Marini-Bettòlo (Rome) 155

Thin-layer chromatography of isomeric oximes. II.
    by I. Pejković-Tadić, M. Hranisavljević-Jakovljević and S. Nešić (Belgrade) 160

The adaptation of the technique of thin-layer chromatography to aminoaciduria investigation
    by J. Opieńska-Blauth, H. Kraczkowski and H. Brzuszkiewicz (Lublin) . . 165

Direct analysis of phospholipids of mitochondria and tissue sections by thin-layer chromatography
  by S. B. Curri, C. R. Rossi and L. Sartorelli (Padova) 174
Qualitative and quantitative analysis of natural and synthetic corticosteroids by thin-layer chromatography
  by G. Cavina and C. Vicari (Rome) 180
Thin-layer chromatography and the detection of stilboestrol
  by C. Bonino (Bologna) 195
The analysis of mixtures of animal and vegetable fats. IV. Separation of sterol acetates by reversed-phase thin-layer chromatography
  by J. W. Copius Peereboom (Leiden) 197
Applications of thin-layer chromatography on Sephadex to the study of proteins
  by P. Fasella, A. Giartosio and C. Turano (Rome) 205
Thin-layer chromatography on silica gel of food colours
  by M. A. Ciasca and C. G. Casinovi (Rome) 212
The application of thin-layer chromatography to investigations of antifermentatives in foodstuffs
  by M. Covello and O. Schettino (Naples) 215
Direkte quantitative Bestimmung von Kationen mittels Dünnschichtchromatographie
  von H. Seiler (Basel) 220
Thin-layer chromatography of inorganic ions. I. Separation of metal dithizonates
  by M. Hranisavljević-Jakovljević, I. Pejković-Tadić (Belgrade) and K. Jakovljević (Slovenia) 221
Index 225

INTRODUCTORY SPEECH

by

Prof. GIORDANO GIACOMELLO

Director, Istituto Superiore di Sanità

It is a great pleasure to me, personally and on behalf of the Istituto Superiore di Sanità, to welcome here today Prof. Cramarossa, who is representing the Minister, and all those taking part in this International Symposium on Thin-Layer Chromatography — the speakers, the various department heads, the many representatives of Italian and foreign universities, the directors of our provincial laboratories and their assistants, and all who are gathered here today to discuss a new analytical technique of great value in basic research and above all in applied research, and which closely concerns several aspects of the activity of this Institute.

On several occasions in the past, the Institute has organized meetings directly or indirectly concerned with problems of health and hygiene.

Last October we held a Symposium on Food Additives to study the orientation and conclusions of scientists of the various European countries concerning the use of chemical additives in foodstuffs. This Symposium was of considerable importance in clarifying many aspects of the subject and in supplying our legislators with a basis upon which to formulate the Law on alimentary substances.

This was followed by a Symposium on Food Microbiology, called to seek the opinion of the most eminent scholars of the subject on one of the points of the Law regarding the hygiene of food substances.

To these we now add today's Symposium on Thin-Layer Chromatography.

The Law of 30 April, 1962, No. 283, provides, in fact, for the elaboration of analytical methods for the identification and the determination of additives in alimentary substances. Thin-layer chromatography today represents one of the most up-to-date techniques of analytical chemistry, combining a remarkable simplicity and rapidity of execution with a specificity and sensitivity greater than can be expected of other chromatographic methods.

This technique, in rapid evolution, today not only offers a refined instrument for the identification of additives and the evaluation of their purity, but also constitutes a highly efficient means for the detection of alimentary frauds.

Considering solely the field of food substances, in the literature already existing on the subject of thin-layer chromatography we have methods for establishing the purity of food dyes — methods which we owe principally to Dr. Wollenweber, here with us today.

We have, furthermore, a series of methods which facilitate the separation and recognition of the main antioxidants and antifermentatives, and I am happy to note that Prof. Covello is to speak to us of his work on food additives.

However, not only additives can be identified with this technique: many substances used as plasticizers can also be isolated, and their identification in very small quantities is of utmost importance in the evaluation of food containers.

Still in the field of alimentation, we must also consider in particular the use of this technique for the separation and identification of lipids and fats, which, in conjunction with gas chromatography, offers the possibility of a complete study of fatty substances.

But the application of thin-layer chromatography is not limited to alimentary analysis. There are many other analytical problems of great interest which it can help us to solve.

Sufficient to mention that it allows a rapid and accurate analysis of insecticides, particularly in agricultural products. In the development of this particular application much has been contributed by Dr. Petrowitz, whom I am happy to be able to welcome among us today. This is a field of traditional interest to the Institute, which devotes to it a great deal of attention under the guidance of Prof. Alessandrini.

Furthermore, thin-layer chromatography permits not only the identification of individual drugs in mixtures, but also the products of decomposition. It is of valuable assistance in the study of antibiotics, and in research for new representatives of this type of compound, particularly when used in conjunction with microbiological techniques.

In the field of toxicology, it permits the separation of alkaloid mixtures and the study of drug metabolism, in that it allows the separation and recognition of the metabolic products of these substances in blood and urine.

Not to be underestimated, furthermore, is its contribution in simplifying the evaluation of the stability of many drugs which are particularly subject to alteration, and it is hardly necessary to mention its value in the examination of vitamin and hormone preparations, etc.

Yet another sphere in which thin-layer chromatography can be of value to us is in the study of atmospheric contamination, when it is necessary to establish the nature of the filter-separated hydrocarbons to identify those which are harmful. Prof. D'Ambrosio and Dr. Pavelka have availed themselves of this technique in their study of atmospheric contamination in Milan.

One might say that there is no field of research to which thin-layer chromatography has not contributed. Even in nuclear studies it has proved its worth, in the separation of radioelements as in the study of the products of radiolysis of the organic moderators of reactors.

However, to remain in the field of major interest to the chemist and biological chemist, mention should be made of the contribution this technique has made and will make to the study of natural substances of animal and vegetable origin, as is also shown by the research carried out in this very Institute under the direction of Prof. Marini-Bettòlo.

Modern pharmacognosy, too — as demonstrated by the research work of Prof. Stahl — has benefited from its use, as has the chemistry of aromatics and essential oils, as Dr. Demole has shown.

In biological chemistry its application to the separation of nucleic acids, of sugars, of amino acids, of steroids, has led to important results, owed particularly to the work of Dr. Neher, which demonstrate, among other things, the possibility of following the metabolic processes of the various constituents. In this same category are Prof. Opieńska-Blauth's studies on urinary amino acids, and those of Prof. Siliprandi and his colleagues on mitochondria.

To conclude this brief summing-up, I would like to emphasize that thin-layer chromatography, initiated with a layer of silica gel, is undergoing constant evolution, as did column chromatography.

Thus we have thin-layer alumina chromatography developed especially by the Czechoslovakian authors, Dr. Lábler among them; the techniques based on the use of cellulose paste; of exchange resins; of polyamides; and finally of dextran derivatives which constitute a novelty of this Symposium, presented by the School of Biological Chemistry of the University of Rome.

To the aspect of qualitative analysis must be added the preparatory aspect, to be illustrated by Drs. Seiler and Fétizon.

In conclusion, I once again extend to each of you here present a very sincere welcome, and would like to move a vote of thanks to Dr. M. Lederer, who, with Prof. Marini-Bettòlo, arranged this very pleasant gathering. It has already proved worthwhile by bringing together in collaboration so many eminent scholars from every country, and I trust that it will prove equally fruitful from every point of view.

DEVELOPMENT AND APPLICATION OF THIN-LAYER CHROMATOGRAPHY*

EGON STAHL

Institute of Pharmacognosy, Saarland University, Saarbrücken-15 (West-Germany)

1. INTRODUCTION

A few years ago the concept of thin-layer chromatography was unknown and only a few "outsiders" were using thin, inorganic sorption layers for separating lipids. Now, tens of thousands of mixtures are analysed daily this way and two books have recently appeared on this subject in Germany. Compared to other methods, which are coming into increasing use and which often demand considerable apparatus and servicing personnel, the method to be described here is basically simple and easy to learn.

Before discussing the general principle of the experimental technique, its applications, and possibilities for future development, here is a "curriculum vitae" of thin-layer chromatography.

2. HISTORY OF DEVELOPMENT

In the thirties there was a persistent search in organic chemistry for an adsorption-chromatographic micro-separation process. Tswett-columns, 2–5 or more cm thick, were replaced by thin capillary tubes. The problem does not lay so much in the separation as in the detection of the substances separated. The decisive step forward was, in my opinion, the change to "open" separation columns, which avoided many of the earlier difficulties and which was first described in 1938 by IZMAILOV AND SCHRAIBER. Swiss colleagues have informed me that Professor IZMAILOV (Fig. 1) died in 1961 while Director of the Pharmaceutical Institute of the University of Kharkov and that, although more than 200 publications came from his pen, he did no more on "drop chromatography" as he then called his method. He took specimen plates on which was spread a 2-mm thick, firmly adhesive layer of alumina, introduced 1 drop of alcoholic plant extract to the layer and gradually dropped alcohol on to the centre of this spot. By this means he obtained small round chromatograms. At the side of Fig. 2, which is taken from his first report, will be seen a series of illustrations. At that time he used the method to investigate all the tinctures of the Russian pharma-

---

* 11th Communication.

Fig. 1. N. A. IZMAILOV,
born June 26th, 1907, died October 2nd, 1961.

copeia and demonstrated the advantages of this procedure over column chromatography. A short while later, CROWE in the U.S.A., referring to this work, reported that he had used loose aluminium oxide layers in the same way as IZMAILOV AND SCHRAIBER, but only, however, as a preliminary test for column chromatography. WILLIAMS also investigated the technique and protected the loose layer by covering with a second glass plate. Similarly in England the CONSDEN, GORDON AND MARTIN group was confronted with the problem of repeating their partition-chromatographic separations of amino acids, first carried out in columns, on a micro-scale. This group also went over to "open" columns, in this case filter paper. As a result of the as-

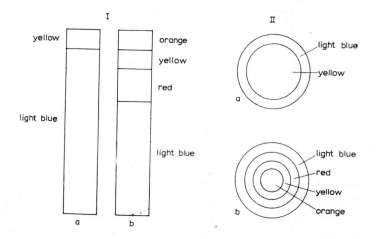

Fig. 2.
Comparison between the fluorescence colours of an alumina column chromatogram (I) and a "drop" chromatogram (II) of Belladonna extract. *a* before developing, *b* after developing with alcohol.

tounding success of this method, all kinds of separation problems were then investigated by this means. New solvents were continually tried and, if no progress was achieved, attempts were made to change the characteristics of the cellulose by impregnation or by chemical transformations. In order to carry out adsorption-chromatographic separations on cellulose- or glass fibre paper in the same way as had been done with columns, the papers were impregnated with alumina or silicic acid. KIRCHNER was one of the first to recommend this method. It is therefore not clear why he, in collaboration with MILLER in 1951, referred back to the *Surface Chromatography* of MEINHARD AND HALL which appeared in 1948. It is, however, certain that this technique was also based on IZMAILOV's separation layers, except that longer strips of glass were used and the name "Chromatostrip Technique" was employed. Despite a series of excellent papers from the group of MILLER, KIRCHNER AND KELLER in the U.S.A., the method was only used occasionally in a few laboratories dealing with terpene derivatives. Perhaps the reason was that the technique was not yet satisfactorily perfected. It is a general rule that a method evolves from the problem, *i.e.* the goal of the investigation, and so it was in our own team. Eleven years ago, and still to-day, we wanted to separate the contents of individual plant and animal cells. This was not possible with any of the existing methods, so we first turned our attention to the structure of the separation layers. The coarse fibre structure of the papers, and also the commercial adsorption media for column chromatography made the very small amounts of substance used (less than 0.1 μg) disappear. So we went on increasingly to finer and thinner sorption layers until we finally managed to separate chromatographically the contents of a few plant-glands, these being practically invisible to the naked eye, on a silica-gel layer 20 μ thick. In this way we showed the products contained therein could in no way be identified with the isolated products. The advantages of such fine grain separation layers became increasingly clear to us during the subsequent years, and we learned how to separate more and more new classes of compounds. Our first publication on the subject, in 1956, passed with as little notice as that of IZMAILOV AND SCHRAIBER and their successors, and we began to wonder why the method was not being generally adopted. At first we had to contend with the objection that it was another of the current chromatographic gimmicks, of the sort we have seen in recent years even in thin-layer chromatography. But, giving up trying to solve individual problems, we devoted ourselves for five years to investigating the method and have tried to evolve the best possible procedures.

By 1958 our work was sufficiently advanced for us to present the method to a large number of scientists in the "ACHEMA 1958" (Frankfurt) and the "ILMAC 1959" (Basle). At first, thin-layer chromatography was taken up by the industrial laboratories of South Germany and Switzerland and from there it has spread to the whole western world. Even in Eastern Europe and the Soviet Union the method is known, but there they work mainly with loose layers, as had been done from 1951 to 1956 particularly by MOTTIER and his colleagues in Switzerland.

3. EXPERIMENTAL TECHNIQUE

The following points seemed important to us:

(*i*) Rational preparation of uniform thin layers.
(*ii*) Arranging the necessary equipment into a single basic apparatus suitable for immediate use.
(*iii*) Finding the most universally applicable sorption media.
(*iv*) Ascertaining the factors influencing the process, and establishing standards, *i.e.* creating a starting basis.
(*v*) Deciding the range of application of the method.

Preparation of thin separation layers

Thin layers may be prepared in several ways and there is little need to scratch one's head over the matter, as such techniques are already known in the film, varnish, and manufacturing industries. A suspension of sorbent may be applied to a fixed carrier-plate, *e.g.* of glass or metal in the following ways:

(*a*) Spreading
(*b*) Pouring
(*c*) Spraying
(*d*) Dipping.

Fig. 3. Preparation of a layer with the adjustable Desaga-TLC-spreader. (Manufactured by Fa. C. Desaga, Heidelberg, W.-Germany.)

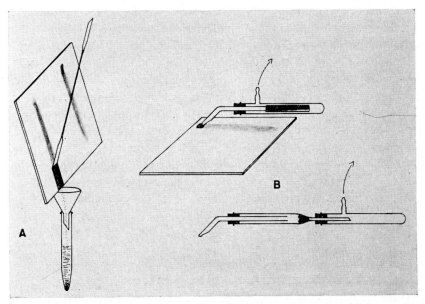

Fig. 4. Isolation of thin-layer chromatographically-separated substances. (A) scratching off, (B) with a simple "vacuum cleaner".

All these methods have been tested by us, insofar as they affect the practical needs of a medium-sized laboratory. In accordance with the stipulation for a rational preparation of layers of defined thickness, the method should also correspond, as regards safety and cleanliness, with the laboratory facilities. The last few years have shown that the following suggested standardisation of the method may be taken as a good basis. Most analytical separation problems may be resolved on a square 20 × 20 cm plate to which is applied a 250-μ thick, adhesive layer. It is also being increasingly recognised that thin-layer chromatography can be an excellent tool for micro-preparative separation of mixtures. For this, it is better to use thicker layers of 0.5 to 1 mm and glass plates 40 cm wide. These allow easy separation of mixtures in amounts of 10 mg up to a maximum of 500 mg, giving sufficient material for determination of chemical composition or of physical constants. The problem of adjusting the layer thickness is easily overcome by use of the thin-layer applicator we have developed (Fig. 3). For micro-preparative separation, however, it has hitherto been extremely difficult to apply a starting band of 1 ml of a solution on to the layer with sufficient uniformity to obtain the optimal separation effect. A small device — recently also described by RITTER AND MEYER — can help us here. The substances are not applied drop by drop to form the starting band, but are sprayed on evenly in a line (see Fig. 11 in *Dünnschicht-Chromatographie, ein Laboratoriumshandbuch*, Springer, Berlin). There is still some difficulty in determining the exact dosage, but we hope that this problem may also be overcome.

Both in micro-preparative work, and in quantitative evaluations of thin-layer chromatograms, there is the task of recovering the substances quantitatively from the plate and of extracting them. Fig. 4 sketches the possibilities available so far. With a sharp spatula, the layer is removed directly into a centrifuge tube filled with the extraction medium; this is then shaken and the sorbent centrifuged off. From the clear residual solution the substance may be obtained or directly determined photometrically. For substances not sensitive to oxidation a micro-"vacuum cleaner", arranged from simple laboratory equipment may be used. We use two types of these micro-"vacuum cleaners". The upper (Fig. 4B) corresponds in principle to that described by RITTER AND MEYER, in the lower a percolation tube serves to gather up the material. By applying solvents the substance may be then directly extracted. This procedure however should not be used for the many oxidation-sensitive natural substances. One should also be very careful when applying the substances to the sorption layers which are often highly active. It is often advisable to carry out application of the samples in a protective atmosphere of inert gas. A small preparation box has been constructed specially for this purpose.

Separation chambers for thin-layer chromatography

At first little attention was paid to the influence of the separation chamber. Even today there is often a tendency to ignore it, as there is to ignore other intermediate influencing factors, and R_F-values are published which cannot be reproduced. In many cases this is certainly not intended, for as soon as one comes from paper chromatography, one is astounded at the smallness of the spots and the much better separation effects and one overlooks the fact that there is still much to be considered. Thus it may be observed that, if solvent mixtures and ordinary aquarium tanks are used, an undesirable edge effect is obtained, as described by DEMOLE and by us in 1958. This stems, we showed, from inadequate saturation of the chamber with the solvent vapour. Fig. 5 demonstrates this. This unsatisfactory saturation of the chamber may however be countered simply by lining the tank completely with filter paper saturated with the solvent. The running times are thereby shortened, and, only if this so-called chamber saturation is employed, can one genuinely speak of R_F-values. It is also possible to reduce the chamber volume to improve saturation.

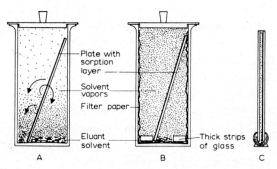

Fig. 5. Separation chambers and saturation. (A) chamber with normal saturation (= NS), (B) chamber saturated with solvent vapours by lining the jar completely with filter paper (= CS), (C) S-chamber.

Fig. 6. BN-chamber for continuous-flow technique. (Manufactured by Fa. C. Desaga, Heidelberg, W.-Germany).

Fig. 7. S-chamber system. The plate is 40 cm wide. (Manufactured by Fa. C. Desaga, Heidelberg, W.-Germany).

Independently and in the contexts of different problems, technical difficulties have been overcome by my colleagues BRENNER AND NIEDERWIESER, using the horizontally arranged "BN-chamber" and by ourselves with the "S-chamber". The next two illustrations will show you these separation-chamber systems specially devised for thin-layer chromatography (Figs. 6 and 7). The BN-chamber, demanding a little more in the way of construction, enables one to use the continuous flow technique, which is of particular importance for the separation of many amino-acid mixtures. Moreover it has the advantage that it is possible to cool it, hence permitting low-temperature chromatography; it is also capable of being used for the ascending technique. In our own laboratory, if we wish to carry out micro-preparations or comparative investigations, *e.g.* if we wish to develop 30 or 40 samples under exactly similar conditions, we use 40-cm wide plates and the S-chamber. In fact these experiments have shown that in many instances the separation effect of the S-chamber is notably better than that obtained from the tank chambers used hitherto. For example, we found that resins and balsams which are very difficult to analyse, spread out into narrow bands, enabling us to recognise many more substances than was possible with the methods used earlier. This finding was confirmed with other mixtures. The central problem of thin-layer chromatography now arises, namely the sorption media.

Sorption media for thin-layer chromatography

In all chromatographic experiments an optimal separation is reached only when the correct *stationary phase* and the appropriate *mobile phase* are used. In paper chromatography one is more or less restricted to cellulose. In column chromatography the dominant adsorbent, in the past as at present, has been aluminium oxide. Usually one attempts to attain separation by varying the mobile phase, that is, by altering the solvent. In thin-layer chromatography we now have the possibility of investigating the most varied kinds of separation layers in a short time and of quickly seeing the separation results. The value of the method is not appreciated if one simply degrades it to a "silica-gel thin-layer chromatography", imagining it to be only a question of taking the coarse-grained sorption media of column chromatography and applying it layer-wise on to a plate for subsequent separation. The progressive step lies in the use of fine-grained material, may it be inorganic or organic. Layers consisting of small particles, 1–25 μ in size, allow many separation problems to be solved in an astonishingly short time. We can only have thin layers of very fine-grained material. Because of this we have taken the step of not calling the method "drop chromatography" or "surface chromatography" or "chromatostrip" or "chromatoplate" or "thin-film chromatography"; rather we name this advance "thin-layer chromatography". In my opinion the most important advance resulting from the method is its ability to render visible uniform separations at an ultra-micro level. This is possible because the concentration of the separated components in each spot area is higher than it is on the fibrous or coarse sorption layers used up to now. Perhaps this can be clarified by the next illustration. In Fig. 8 the zone sizes of a paper chromatogram are compared with those of a thin-layer chromatogram for increasing amounts of substance. Particularly impressive are the comparisons in the work of BRENNER and his associates (Chapter : "Amino Acids and Derivatives" in *Dünnschicht-Chromatographie, ein Laboratoriumshandbuch,* Springer Verlag, Berlin, Göttingen, Heidelberg).

These most important findings can be explained if we consider the grain size and

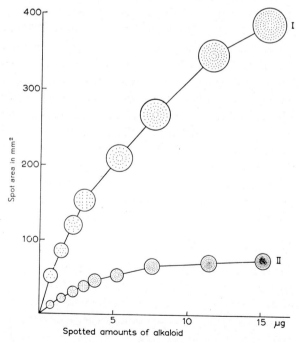

Fig. 8. Comparison between the spot area of separated alkaloids (spotted in increasing amounts) on paper impregnated with formamide (I) and on a silica gel layer (II). Dots symbolise the number of molecules.

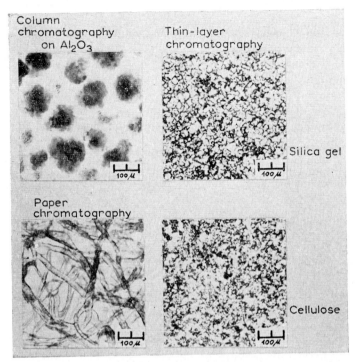

Fig. 9. Comparison between structural differences in sorbents for column chromatography and paper chromatography (left) and in sorbents for thin-layer chromatography (right).

structure of the sorption medium used hitherto, and compare it with those used in thin-layer chromatography (Fig. 9).

If thin-layer chromatography has become world-famous in a few years it is largely due to the Silica gel G specially prepared for this purpose. About 80–90% of all the thin-layer chromatograms described to date were obtained on these layers. In 1956 we did in fact work with pure silica-gel layers without a binding medium, but it became clear that for this kind of work, it would be advantageous to have a more cohesive layer. About 10% gypsum was therefore added, which when the layer was dried, crystallised into matted needles, giving a much better cohesiveness than was obtained in its absence. The calcium sulphate interferes with only a few separation problems although in the meantime other silica gels have been developed which guarantee good cohesiveness even without gypsum addition. This product was designated Silica gel H, where H stands for the property of being "adhesive" (in German: haftend). Although it was a good thing in the first place to find the universally applicable silica gel, which will always be of great value in the future, it should not be forgotten that the method will not be completely developed until the most varied kinds of separation layer have been tried. Optically active silica gels seem to me particularly attractive and capable of development; these were first described years ago in England by BECKETT AND ANDERSON for use in column chromatography. From Table 1 we can see that today there is available for thin-film chromatography a rich

TABLE 1

SORBENTS FOR THIN-LAYER CHROMATOGRAPHY

Inorganic

Silica gel G Merck	since 1958
Silica gel GF_{254} Merck	since 1962
Silica gel H Merck	since 1962
Silica gel HF_{254} Merck	since 1962
Silica gel D-5 Fluka	since 1962
Silica gel D-O Fluka	since 1962
Silica gel D-5-F Fluka	since 1962
Silica gel Woelm	since 1962
Serva-silicagel–TLC	since 1962
Aluminum oxide G Merck	since 1959
Aluminum oxide D-5 Fluka	since 1959
Aluminum oxide D-O Fluka	since 1959
Aluminum oxide D-5-O Fluka	since 1959
Aluminum oxide basic, acid, neutral Woelm	since 1962
Kieselguhr G Merck	since 1959
Magnesium silicate Woelm	since 1962

sec.-Magnesium phosphate
Calcium hydroxide, calcium sulphate
Ferric hydroxides
Glass powder and other silicates
Charcoal

Organic

Cellulose powder MN 300 (and MN 300 G)	Macherey, Nagel and Co.	since 1961
Cellulose powder MN 300 Ac (MN 300 G/Ac)	Macherey, Nagel and Co.	since 1961
Cellulose powder MN 300 CM (MN 300 G/CM)	Macherey, Nagel and Co.	since 1961
Cellulose powder MN 300 P (MN 300 G/P)	Macherey, Nagel and Co.	since 1961
Cellulose powder MN 300 F_{254} (MN 300 GF_{254})	Macherey, Nagel and Co.	since 1963
Cellulose powder MN 300 DEAE (MN 300 G/DEAE)	Macherey, Nagel and Co.	since 1961
Cellulose powder MN 300 ECTEOLA (MN 300 G/ECTEOLA)	Macherey, Nagel and Co.	since 1961
Excorna-cellulose powder C 1000		since 1960
Cellulose powder Schleicher and Schüll		since 1962
Serva-cellulose–TLC		since 1962
Camag-cellulose powder D		since 1962
Camag-cellulose powder DF		since 1962
Sephadex, Pharmacia		since 1962
DEAE-Sephadex, Pharmacia		since 1962
Polyamide powder Merck		since 1962
Polyamide powder Woelm		since 1962

Polyacrylonitrile
Sugar, starch, inulin
Carbamide

palette of inorganic and organic sorption media from which to make a selection. The separation properties of the layer can also be varied by using mixtures of these sorption media: for instance, a mixture of Silica gel G, which is itself acidic, with the basic Alumina G, has been shown to be particularly advantageous. Mixed layers are also suitable for many carotene separations. A whole array of separation problems may be solved by the use of so-called reaction layers. Their preparation is simple: the suspension of the sorption medium is mixed, not with water, but with a dilute acid, base or buffer solution. Complex-forming layers may be obtained, e.g. with boric acid solution or complex-forming heavy metals and salts. Of particular interest for separating *cis–trans*-isomeric lipids are the silver nitrate layers as used by DE VRIES and by MORRIS and others. I consider it an important advance that, for two-dimensional experiments, we can obtain a separation first by adsorption chromatography and then obtain further resolution by partition chromatography after making the layer hydrophobic. Equally interesting and useful is the combination of adsorption thin-layer chromatography with thin-layer ionophoresis.

But these are only a few possibilities which have been selected to show the possibilities of development in the method in this field. I have not yet mentioned the luminescent layers, made known by GÄNSHIRT, which permit observation of the separated compound without disturbing the layer. In short wave U.V.-light of about 254 mμ they give an intense green fluorescence and compounds which absorb in that region are easily visible as dark coloured spots. Silica gels and cellulose powder with added fluorescence indicators have since become commercially available.

This brief introduction may serve to show what a large field still awaits investigation. Systematic exploration is certainly more attractive and of more general interest than trivial investigations into methods for preparing layers. Attempts to create thin layers by pressing or fusing the sorbent together appear to me equally senseless. We thereby only narrow the possible applications of the method. For example, we ourselves 12 years ago used magnesia rods and magnesia grooves with anodic oxidised aluminium foil and etched glass and could only obtain partial success.

4. QUANTITATIVE EVALUATION

Apart from the choice of sorption layers, quantitative evaluation will certainly be more amenable to detailed investigation. Particularly interesting is direct spectrophotometric estimation of the chromatograms. This would be the most attractive method in many cases. Very interesting is the recently devised TLC-scanner, which permits fully automatic detection of radioactively labelled substances after separation by thin-layer chromatography (Fig. 10).

5. APPLICATION OF THIN-LAYER CHROMATOGRAPHY

A clear picture of the widespread use of the method and of its areas of greatest application is obtained by an analysis of publications to date. Fig. 11 shows first that in 1958 and 1959 about 20 papers appeared, and in the following year over 60; 1961 had 150 and in 1962 there were 300 publications. From the graphic representation of the areas of application of the method so far, we see that the greatest utilisation lies in the sphere of lipophilic materials: a similar analysis for paper chromatography would probably show the same for hydrophilic mixtures. As it has been shown that fine-grained cellulose powder has considerable advantages over paper, the next few years will probably reveal a definite trend in favour of mixtures of hydrophilic substances. However we should not overlook the fact that approximately two-thirds or more of all industrially prepared synthetic products possess lipophilic characteristics.

It should be remembered next that nearly all the synthesis steps from carbon to the end-product must be controlled while in progress, and further that this step-wise transformation of the material usually leads to changes in its polarity. Moreover, we are always dealing with reaction mixtures about whose composition we must be quite clear. Thin-layer chromatography is an ideal indicator for rapid recognition of polarity differences, so it will be applied to this field even more than hitherto. I can imagine that it will become a tool as indispensable in the laboratory as are the flasks in which the reactions are carried out. From these a sample is taken from time to time and rapidly chromatographed. From the positions of the resulting spots, depending on their polarity, one can verify that the reaction is progressing correctly and, from the relative sizes of the spots, the extent to which the reaction has proceeded can be determined. Anyone who has once used this convenient method will recognise its invaluable assistance and not wish to abandon it.

In this introduction, however, I should in no way wish to belittle the importance

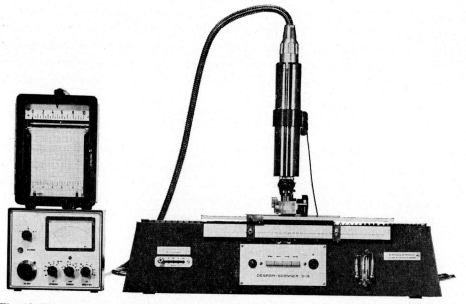

Fig. 10. Desaga-scanner. Fully automatic measurement of all radioactively labelled substances (^{3}H, ^{14}C, ^{35}S, ^{32}P, etc.) after TLC separation. (Photograph Desaga, Heidelberg.)

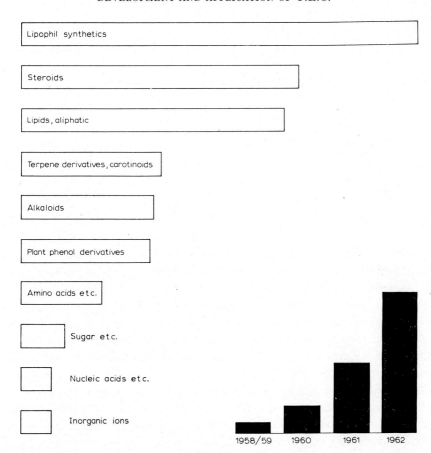

Fig. 11.
Growth of publications on TLC (black areas). Classification of fields of application (white areas).

of the method in the field of the chemistry of natural products; for chromatographic methods are naturally applied there also, and the significance of thin-layer chromatography has been immediately recognised there and cleverly put to use. I should like to mention an important field of application for the future, namely its potentialities in medical research and diagnosis. The detection sensitivity of compounds which are not very prone to decomposition renders possible direct investigation of the body fluids.

But finally, despite all the optimism, a warning against overvaluation of thin-layer chromatography. Only a combination of various types of chromatographic methods, supplemented by classical chemical and physical identification methods, will guard us from serious error.

ACKNOWLEDGEMENT

I thank Dr. H. R. BOLLIGER for providing the photograph of and information about Prof. Dr. IZMAILOV, and I am grateful to Prof. Dr. M. ASHWORTH for his help in translating and proof-reading this article.

DÜNNSCHICHTCHROMATOGRAPHIE AUF CELLULOSE-SCHICHTEN

PAUL WOLLENWEBER

Wiss. Abteilung der Firma Macherey, Nagel & Co., Düren (Deutschland)

1. EINLEITUNG

Im Rahmen dieses Vortrages soll nicht auf die geschichtliche Entwicklung der Dünnschichtchromatographie (DC) eingegangen werden; jedoch soll herausgestellt werden, dass die Anwendung der DC, insbesondere die Anwendung von anorganischen Trägermaterialien nicht zuletzt aus dem Grunde eine so rasche und starke Verbreitung gefunden hat, weil es zur Trennung von lipophilen Substanzgemischen keine schnelle und empfindliche Analysenmethode gab. Auch mit Hilfe der Papierchromatographie (PC) konnten die auf diesem Gebiete anfallenden Probleme nicht alle in zufriedenstellendem Masse gelöst werden. Zu der Schnelligkeit und hohen Empfindlichkeit der dünnschichtchromatographischen (dc-) Methode kam bei Anwendung von anorganischen Schichten der glückliche Umstand hinzu, dass infolge der Anwendbarkeit stark aggressiver Reagenzien der Nachweis von schwierig nachzuweisenden Substanzen vereinfacht oder überhaupt erst möglich wurde. Es ist also eine Folge der Forderung nach einer eleganten Methode zur Trennung lipophiler Substanzgemische, dass zunächst die anorganischen Sorptionsmittel wie Kieselgel, Kieselgur und Aluminiumoxid weite Verbreitung als Trennschichten gefunden haben. Obschon man bei Cellulose-Schichten nicht die aggressiven Nachweisreagenzien anwenden kann wie bei den anorganischen Schichten, lag es doch sehr nahe, auch Cellulose bzw. Cellulosederivate für diese neue Methode einzusetzen, um dem Analytiker weitere Möglichkeiten der Chromatographie auf dünnen Schichten für die vielfältigen Probleme in die Hand zu geben. Es stellte sich nämlich bald heraus, dass auch die Chromatographie auf Cellulose-Schichten viele Vorteile gegenüber der Chromatographie auf Papier bot und dass sie eine wertvolle Ergänzung der DC auf anorganischen Schichten darstellte.

Aufbauend auf den Erfahrungen und Kenntnissen zur Herstellung von Papieren und Cellulosepulvern für die Papier- und Säulenchromatographie wurden in Zusammenarbeit mit Herrn Prof. STAHL zur ACHEMA-Tagung 1961 zunächst das normale Cellulosepulver mit und ohne Gipszusatz in standardisierter Form dem Analytiker zugänglich gemacht. Inzwischen steht ein umfangreiches Sortiment Cellulosepulver mit und ohne Bindemittel standardisiert nach E. STAHL für die Dünnschichtchromatographie zur Verfügung.

Es handelt sich um normale, unbehandelte Cellulose, um normale Cellulose mit

TABELLE 1
MN-CELLULOSEPULVER FÜR DIE DÜNNSCHICHTCHROMATOGRAPHIE

	Cellulose-Austauscher	Kapazität (mÄq/g)
	Sorten ohne Bindemittel	
MN 300	Normale Cellulose	—
MN 300 F$_{254}$	MN 300 mit Leuchtstoff	—
MN 300 Ac	Acetylierte Cellulose (ca. 10%)	—
MN 300 CM*	Carboxymethyl-Cellulose	ca. 0,7
MN 300 P*	Phosphorylierte Cellulose	ca. 0,7
MN 300 DEAE**	Diäthylaminoäthyl-Cellulose	ca. 0,7
MN 300 ECTEOLA**	Cellulose-Derivat (Struktur liegt nicht fest)	ca. 0,35
MN 300 PEI**	Polyäthylenimin-imprägnierte Cellulose	ca. 0,7
MN 300 Poly-P*	Polyphosphat-imprägnierte Cellulose	ca. 0,7
	Sorten mit Gipszusatz	
MN 300 G	Normale Cellulose	—
MN 300 GF$_{254}$	MN 300 G mit Leuchtstoff	—
MN 300 G/Ac	Acetylierte Cellulose (ca. 10%)	—
MN 300 G/CM*	Carboxymethyl-Cellulose	ca. 0,7
MN 300 G/P*	Phosphorylierte Cellulose	ca. 0,7
MN 300 G/DEAE**	Diäthylaminoäthyl-Cellulose	ca. 0,7
MN 300 G/ECTEOLA**	Cellulose-Derivat (Struktur liegt nicht fest)	ca. 0,35

* Kationen-Austauscher, ** Anionen-Austauscher.

Leuchtstoffzusatz, um acetylierte Cellulose und um die aus der Säulenchromatographie bekannten 4 Typen Cellulose-Ionenaustauscher:

	aktive Gruppen
Carboxymethyl-Cellulose (CM)	$R-O-CH_2COOH$
Phosphorylierte Cellulose (P)	$R-O-PO_3H_2$
Diäthylaminoäthyl-Cellulose (DEAE)	$R-O-C_2H_4N(C_2H_5)_2$
ECTEOLA-Cellulose	Nicht eindeutig bekannt

Letztere Sorte ist ein Reaktionsprodukt aus Alkali-Cellulose, Triäthanolamin und Epichlorhydrin. Die genaue Struktur liegt nicht fest.

Bevor ich auf die einzelnen Typen und ihre Anwendung eingehe, möchte ich noch kurz einige allgemeine Betrachtungen vorausschicken. Wie bei Kieselgel-Schichten zeigen sich auch bei Cellulose-Schichten grundsätzlich dieselben Vorteile gegenüber der Papierchromatographie, nämlich:

(1) erheblich höhere Empfindlichkeit der Trennungen auf Grund der geringeren Ausbreitung der Substanzflecken,
(2) wesentlich kürzere Laufstrecken,
(3) als Folge von (2) wesentlich kürzere Entwicklungszeiten.

Die Anwendung stark aggressiver Nachweisreagenzien wie konz. Schwefelsäure ist — wie bereits erwähnt — bei Cellulose-Schichten naturgemäss nicht möglich.

Literatur S. 30

Abb. 1a. Mikroaufnahme von Cellulosepulver für die Dünnschichtchromatographie (Vergrösserung 125-fach).

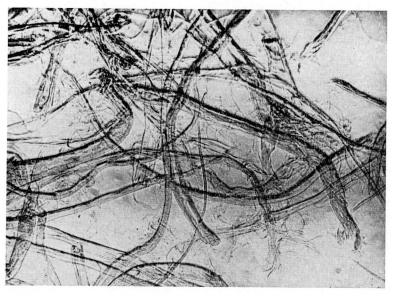

Abb. 1b. Mikroaufnahme von Cellulosefasern aus Chromatographiepapier (Vergrösserung 125-fach)

Bei der Chromatographie auf Schichten aus Cellulosepulver können weitgehend die in der Papierchromatographie üblichen Bedingungen übernommen werden. Daraus kann man folgern, dass der Trennvorgang auf Schichten aus normalem Cellulosepulver auf Verteilung beruht. Die gut erhaltene Faserstruktur der Chromatographiepapiere ist beim Cellulosepulver für die DC nicht mehr vorhanden (vgl. Abb. 1). Dies hat zur Folge, dass blitzartige Substanzausbreitungen entlang der Grenzflächen von langen Fasern nicht mehr möglich sind und dass infolgedessen die Substanzflecken bei gleicher Konzentration kompakter sind als auf Papier. Weiterhin ist die spezifische Oberfläche von Dünnschicht-Cellulosepulvern sehr gross, so dass auch dadurch die Substanzaufnahme auf kleinem Raum begünstigt wird. Als weiteren Vorteil der Pulver gegenüber Papier kann man die Tatsache ansehen, dass Verunreinigungen und Begleitstoffe, die z.B. quantitative Auswertungen stören könnten, viel leichter aus Pulver entfernt werden können als aus Papier.

Als ein für die Praxis sehr willkommener Vorteil der DC auf Cellulose-Schichten gegenüber der Chromatographie auf anorganischen Schichten ohne Bindemittel oder mit Gipszusatz muss die ausserordentlich gute Haftfestigkeit von Cellulose-Schichten lobend hervorgehoben werden. Darüber dürfte vor allem der Praktiker sehr erfreut sein, dem die enorme Abriebempfindlichkeit von Kieselgel-Schichten sicherlich schon oft Verdruss bereitet hat. Die Haftfestigkeit von MN-Cellulose-Schichten ist nur in verschwindend geringem Masse vom zugesetzten Bindemittel abhängig. Cellulose-Schichten ohne Bindemittel haften nämlich ebenfalls sehr fest auf der Platte. Auf die Frage, warum setzt man denn dem Cellulosepulver überhaupt Gips zu, kann geantwortet werden, dass die Praxis lehrte, dass Schichten mit und ohne Bindemittel unterschiedliche Trenneffekte zeigen können. Ähnliche Erscheinungen hat der aufmerksame Beobachter auch schon bei Kieselgel mit und ohne Bindemittel feststellen können. Die Tendenz geht jedoch langsam dahin, dass sich normales Cellulosepulver ohne Bindemittel, also die Sorte MN 300, weit öfter mit gutem Erfolg einsetzen lässt als die entsprechende Sorte MN 300 G mit Bindemittel. Bei Ionenaustauscher-Cellulosepulvern liegen die Verhältnisse etwas anders. Schichten aus reinem Austauscherpulver haften zwar auch sehr fest auf der Platte, aber sie bekommen beim Trocknen infolge geringer Quell- und Schrumpfvorgänge leicht feine Risse und bilden keine ideale geschlossene Schicht. Diese nachteilige Erscheinung kann man abstellen, indem man den Austauscherpulvern entweder Gips oder aber besser normales Cellulosepulver MN 300 in geringen Mengen beimischt. Lediglich acetyliertes Cellulosepulver liefert Schichten von geringerer Haftfestigkeit, entspricht aber immer noch der Haftfestigkeit des Kieselgel G.

Die gute Haftfestigkeit der Cellulose-Schichten auf der Glasplatte hat zur Folge, dass das Abziehen der Schichten nach Besprühen mit einem Kunstharz für Dokumentationszwecke Schwierigkeiten bereitet. Mit dem jetzt im Handel erhältlichen „Konservierungslack Fluka für Dünnschichtchromatogramme" lassen sich — wenn auch auf umständliche Weise — Cellulose-Schichten von der Platte abziehen. Alle übrigen bisher beschriebenen Methoden und angebotenen Kunstharze bleiben bei Cellulose-Schichten ohne Erfolg.

Literatur S. 30

2. EXPERIMENTELLES

Zur Herstellung von Cellulose-Schichten bedient man sich zweckmässigerweise der im Handel befindlichen Streichgeräte. Wir wenden mit sehr gutem Erfolg die Desaga-Streichgeräte mit und ohne verstellbare Schichtdicke an. Die Dicken der mit letzterem Gerät hergestellten trockenen Cellulose-Schichten liegen in dem unteren Bereich der Dicken von Chromatographiepapieren, nämlich bei etwa 0,12–0,15 mm. Diese Dicke entspricht einem Austrittsschlitz am Streichgerät von etwa 0,25 mm. Schichten dieser Dicke sind für die meisten Trennungen besonders vorteilhaft und offenbaren am sinnfälligsten die Vorteile der DC, d.h. der Chromatographie auf „dünnen" Schichten.

Für präparative Trennungen kann man mit geeigneten Streichgeräten Cellulose-Schichten bis zu einer Dicke von etwa 0,5 mm im trockenen Zustand ohne Schwierigkeiten herstellen. Diese Dicke entspricht einem Austrittsschlitz von etwa 1 mm. Die Herstellung von Schichten von noch grösserer Dicke in einem einzigen Streichvorgang bereitet Schwierigkeiten und man verfährt am besten so, dass man mehrere dünnere Schichten nach jeweiliger Zwischentrocknung übereinander streicht. Es ist sehr erfreulich, dass selbst die dicken Schichten aus MN-Cellulosepulver sehr fest auf der Platte haften und sich leicht handhaben lassen.

Zur Herstellung glatter Cellulose-Schichten sind zwei Punkte besonders zu beachten:

(1) Die Glasplatten müssen völlig frei sein von fettartigen Verunreinigungen.
(2) Alle Cellulosepulver-Suspensionen, d.h. sowohl normale Cellulose als auch Cellulosederivate, müssen intensiv homogenisiert werden.

Die Reinigung der Glasplatten kann auf einfache Weise erfolgen mittels konzentrierter Sodalösung und zwar lässt man die Platten während der Zeit, in der sie nicht benötigt werden, in dieser Lösung liegen. Vor Gebrauch werden die Platten kurz mit Wasser klargespült und getrocknet.

Die Homogenisierung der Cellulosepulver-Aufschlämmungen ist im Gegensatz zu den anorganischen Sorptionsmitteln in einem elektrischen Mixgerät vorzunehmen. Ohne eine solche intensive Behandlung der Pulver-Aufschlämmungen bekommt man keine glatten Schichten. Wir benutzen mit bestem Erfolg das in der Bundesrepublik als Starmix bekannte Haushaltmixgerät, es genügt aber jeder elektrische Rührstab.

Ein weiterer Punkt, der bei der Herstellung von Cellulose-Schichten von genau gleicher Schichtdicke beachtet werden muss, ist die Konsistenz der Cellulose-Suspensionen. Diese dürfen weder zu dick- noch zu dünnflüssig sein. Gegebenenfalls sind die Anrührvorschriften in dieser Hinsicht zu ändern.

Die Trocknungsbedingungen der Schichten sind je nach Pulversorte verschieden. Folgende Trocknungsbedingungen haben sich bewährt: Schichten aus normalem Cellulosepulver werden bei 105° C während 10 Minuten getrocknet, Schichten aus acetyliertem Cellulosepulver am besten an der Luft und Schichten aus Ionenaustauscherpulver bei maximal 50° C während etwa 40 Minuten. Generell kann empfohlen werden, die Platten zunächst an der Luft leicht antrocknen zu lassen.

3. SPEZIELLER TEIL

Allgemeine Anwendungen

Bei der Wahl des Sorptionsmittels wird man von den Eigenschaften der zu trennenden Substanzen ausgehen. Das wichtigste Kriterium ist die Löslichkeit der zu trennenden Stoffe, weil die Chromatographie auf Cellulose-Schichten analog derjenigen auf Papier weitgehend auf Verteilung beruht. Man unterscheidet im wesentlichen die beiden Hauptgruppen hydrophile und hydrophobe (lipophile) Substanzen.

Nicht alle Verbindungen lassen sich streng in dieses Schema einordnen; es gibt vielmehr eine Anzahl Übergänge zwischen diesen beiden Hauptgruppen. Infolge dieser unterschiedlichen Übergänge kann man nicht mit Sicherheit voraussagen, welche Substanzen sich einwandfrei auf normalen Cellulose-Schichten trennen lassen. Mit Sicherheit lassen sich bei richtiger Wahl des Laufmittels hydrophile Substanzen trennen.

Trennungen von rein lipophilen Substanzen sind auch auf Cellulose-Schichten möglich, jedoch bedürfen die Schichten dann einer Vorbehandlung. Diese Vorbehandlung der Schichten kann analog den Methoden in der Papierchromatographie durchgeführt werden und zwar durch Tauchimprägnierung mit Formamid, Dimethylformamid, Paraffinöl etc. Durch solche Imprägnierungen bewirkt man auch eine Umkehrung der Phasen, (bei Paraffinöl) wie sie als „reversed-phase chromatography" in der Papierchromatographie bekannt sind. Die etwas unbequeme Vorbehandlung der Schichten kann umgangen werden durch den Einsatz von acetyliertem Cellulosepulver. Durch solche Manipulationen können grundsätzlich alle anfallenden Trennprobleme auf Cellulose-Schichten bewältigt werden.

Jedoch können Schwierigkeiten dadurch auftreten, dass die getrennten Substanzen entweder sehr schlecht oder überhaupt nicht nachgewiesen werden können. In solchen Fällen wird man der anorganischen Trennschicht den Vorzug geben. In der Praxis verfährt man am besten so, dass man in Vorversuchen sowohl Cellulose- als auch anorganische Schichten zur Lösung eines bestimmten Trennproblems einsetzt und nach den praktischen Ergebnissen entscheidet: welches ist das geeignete Trennmedium?

Ergänzend zu den Ausführungen über die verschiedenen Cellulosepulversorten möchte ich erwähnen, dass das Cellulosepulver mit Leuchtstoffzusatz, also die Sorten MN 300 F_{254} und MN 300 GF_{254}, nicht uneingeschränkt für alle Laufmittel eingesetzt werden kann. Es hat sich gezeigt, dass komplexbildende Substanzen wie etwa Pyridin oder Oxalsäure die Fluoreszenz löschen. Als Leuchtstoff findet ein anorganisches Leuchtpigment Verwendung, das im UV-Licht bei einer Wellenlänge von 254 mμ das Absorptionsmaximum hat. Daher stammt die Zahl 254 als Index bei der Sortenbezeichnung. Der Einsatz von Cellulose-Schichten mit Leuchtstoff erfolgt bei solchen Substanzen, deren Nachweis mittels kurzwelligem UV-Licht möglich ist. Praktisch sichtbar werden die getrennten Substanzen durch eine Fluoreszenzlöschung, d.h. es entstehen dunkle Flecken auf stark grün fluoreszierendem Untergrund. Cellulose-Schichten mit Leuchtstoffzusatz haben den Vorteil, dass man während der

Literatur S. 30

Entwicklung des Chromatogramms den Stand der Trennung beobachten kann. Cellulose-Ionenaustauscher besitzen ionenaustauschende Eigenschaften wie die Kunstharz-Ionenaustauscher; sie unterscheiden sich aber von diesen in einigen wichtigen Punkten. So ist das Cellulosegerüst hydrophil im Gegensatz zur hydrophoben Kunstharzmatrix. Die Mehrzahl der austauschaktiven Gruppen liegen bei der Cellulose nahe an der Oberfläche. Dadurch können sie auch für den Austausch grosser Moleküle wirksam werden, die normalerweise in die Poren eines Kunstharzaustauschers nicht eindringen können. Dies führt zu einem raschen Austausch und dazu, dass Cellulose-Ionenaustauscher trotz einer zahlenmässig weit kleineren Austauschkapazität eine grössere Kapazität für grosse Moleküle haben als Kunstharz-Austauscher. Die Kapazitäten für Cellulose-Austauscher liegen in der Grössenordnung von 0,7 mÄq/g bei den Sorten Carboxymethyl-, phosphorylierte und DEAE-Cellulose und bei etwa 0,35 mÄq/g bei dem Anionenaustauscher ECTEOLA. Für eine gute Reproduzierbarkeit von Trennungen ist es wichtig, dass die aktiven Gruppen auch wirklich in der austauschaktivsten Form vorliegen. Erst dann kommt die ionenaustauschende Wirkung voll zur Geltung. Das „Aktivmachen" geht in der Weise vor sich, dass man die Kationenaustauscher CM- und P-Cellulose mit 0,5 N Salzsäure und die Anionenaustauscher DEAE- und ECTEOLA-Cellulose mit 0,5 N Natronlauge aufschlämmt, etwa 30 Minuten stehen lässt, filtriert und mit destilliertem Wasser weitgehend neutral wäscht. Zweckmässigerweise nimmt man das Regenerieren unmittelbar vor der Herstellung der Schichten vor und schlämmt das noch feuchte Pulver mit der notwendigen Menge Wasser zur Herstellung der Schichten auf. Vollständiges Trocknen des Pulvers ohne nachträgliche schonende Vermahlung führt zu Schwierigkeiten bei der Herstellung der Schichten. Als Laufmittel für Cellulose-Austauscher-Schichten haben sich bewährt: wässrige Pufferlösungen verschiedener Stärke, schwache Normallösungen von Säuren und Laugen sowie Kochsalzlösungen verschiedener Konzentration.

Spezielle Anwendungen

Infolge des rein zeitlichen Vorsprungs der DC auf anorganischen Schichten ist die Zahl der Publikationen von dc-Trennungen auf Cellulose-Schichten noch gering. Jedoch wird sich die Anzahl solcher Publikationen in diesem Jahre stark erhöhen. Über die bis etwa Ende 1962 von uns erfassten Publikationen wird in den folgenden Kapiteln kurz berichtet. Mit gutem Erfolg konnten getrennt werden: Alkaloide[1-3], aliphatische Amine[1], Aminosäuren[1,4,5], Farbstoffe[6-9], Metallkationen[10], Tricarbonsäurezyklus-Substrate[11], Zucker[12,13], Uronsäuren[12,13], Quellstoffe[13], phenolische Verbindungen[15], Nucleinsäurederivate[16-21] und polycyclische Kohlenwasserstoffe[22,23].

TEICHERT, MUTSCHLER UND ROCHELMEYER haben sich zuerst mit der Chromatographie auf Cellulose-Schichten beschäftigt. Sie beschreiben Trennungen von wasserunlöslichen Mutterkornalkaloiden[1] auf mit Formamid imprägnierten Cellulose-Schichten nach erfolglosen Trennversuchen auf Kieselgel-Schichten. Die Imprägnierung der Schichten erfolgt durch Eintauchen in eine 20%ige Lösung von Formamid in Aceton und anschliessendes Trocknen unter dem Kaltluftventilator während 5 Minuten. Die Trennschärfe konnte durch zweimalige Entwicklung in derselben Lauf-

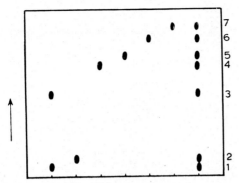

Abb. 2. Trennung von wasserunlöslichen Mutterkornalkaloiden auf formamid-imprägnierten Cellulosepulverplatten mit Benzol/Heptan/Chloroform (6 : 5 : 3) als erstem Laufmittel und Benzol/Heptan (6 : 5) als zweitem Laufmittel. Sichtbarmachung durch gefiltertes UV-Licht: 1 = Ergotamin, 2 = Ergosin, 3 = Ergotaminin, 4 = Ergocristin, 5 = Ergocornin, 6 = Ergokryptin, 7 = Ergocristinin.

richtung erheblich verbessert werden (vgl. Abb. 2). In ca. 75 Minuten können auf diese Weise die Ergotamin- und die Ergotoxingruppe getrennt werden. Als Laufmittel verwendeten sie Gemische von Benzol, Heptan und Chloroform. Der Nachweis erfolgte mittels UV-Licht. Die wasserlöslichen Mutterkornalkaloide bleiben am Start.

Ebenfalls auf formamid-imprägnierten Cellulose-Schichten und unter ähnlichen Bedingungen trennten TEICHERT et al.[2,3] weitere Alkaloidgruppen, z.B. Belladonna-Alkaloide, Rauwolfia-Alkaloide, Opium-Alkaloide, Lobelia- und Chelidonium-Alkaloide. Die Nachweisgrenze liegt nach Abtreiben des Formamids bei 1 γ. Das Abtreiben des Formamids erfolgt während 15 Minuten bei 110° C im Vakuumtrockenschrank bei etwa 20 Torr.

Folgende aliphatischen Amine konnten mit gutem Erfolg auf normalen Cellulose-Schichten mit dem Laufmittel Amylalkohol : Eisessig : Wasser (4 : 1 : 5) getrennt werden[1]:

Methylamin
Äthylamin
n-Propylamin
n-Butylamin
Isoamylamin
Äthanolamin
Histamin
Tyramin.

In einer der bereits genannten Arbeiten[1] weisen TEICHERT, MUTSCHLER UND ROCHELMEYER kurz darauf hin, dass sich Aminosäuren auf normalen Cellulose-Schichten mit dem Laufmittel n-Butanol : Eisessig : Wasser (4 : 1 : 5, obere Phase) trennen lassen. In unserem Laboratorium wurden die Trennmöglichkeiten von Aminosäuren auf Cellulose-Schichten näher untersucht und zwar unter Anwendung von 9 verschiedenen Laufmitteln[4] (vgl. Tabelle 2). Das Ergebnis dieser Untersuchungen lässt sich wie folgt

Literatur S. 30

TABELLE 2
ZUSAMMENSTELLUNG DER UNTERSUCHTEN LAUFMITTEL

L_1 = n-Butanol–Eisessig–Wasser (4 : 1 : 5), obere Phase
L_2 = Pyridin–Methyläthylketon–Wasser (15 : 70 : 15)
L_3 = Propanol–Wasser (7 : 3)
L_4 = Methanol–Wasser–Pyridin (80 : 20 : 4)
L_5 = n-Butanol–Ameisensäure–Wasser (75 : 15 : 10)
L_6 = Propanol–8.8%iges Ammoniak (8 : 2)
L_7 = Äthanol–n-Butanol–Wasser–Propionsäure (10 : 10 : 5 : 2)
L_8 = Phenol–Wasser (8 : 2) (nicht geeignet!)
L_9 = Phenol, mit Phosphatpufferlösung pH 12 gesättigt;
Pufferlösung: 0.067 M NaOH + 0.67 M Na$_2$HPO$_4$. 12 H$_2$O (1 : 1) (nicht geeignet!).

zusammenfassen: die besten Trennungen wurden erzielt mit dem Laufmittel n-Butanol : Eisessig : Wasser (4 : 1 : 5, obere Phase) und zwar auf Schichten aus Cellulosepulver MN 300. Cellulose-Schichten mit Gipszusatz erwiesen sich als völlig unbrauchbar. Auch mit Hilfe der DC können nicht alle Aminosäuren in einem eindimensionalen

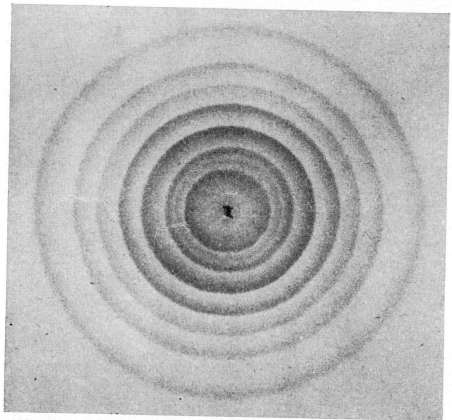

Abb. 3. Horizontal entwickeltes Cellulose-Dünnschichtchromatogramm von 8 synthetischen Lebensmittelfarbstoffen.

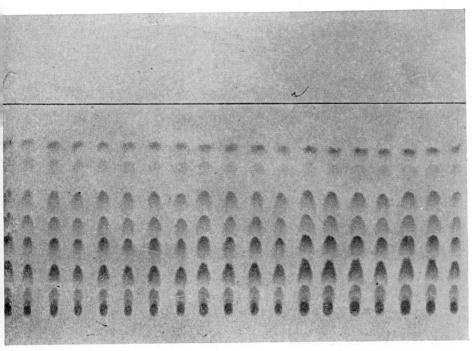

Abb. 4. Cellulose-Dünnschichtchromatogramm von 8 synthetischen Lebensmittelfarbstoffen.

Chromatogramm getrennt werden. Die Nachweisgrenze liegt noch unter 1 γ und dürfte damit niedriger sein als auf Papier. Die Entwicklungsdauer beträgt 30–60 Minuten; der Nachweis erfolgt mit Ninhydrin. Inzwischen wurde mir bekannt, dass man durch geschickte Kombination von verschiedenen Laufmitteln in einem System von 3 Cellulose-Platten der Grösse 20 × 20 cm über 50 verschiedene Aminosäuren sicher identifizieren konnte. Eine entsprechende Publikation befindet sich im Druck[5].

Farbstoffe können auch mit gutem Erfolg auf Cellulose-Schichten getrennt werden. Synthetische Lebensmittelfarbstoffe[6,8,9] können mit verschiedenen Laufmitteln auf Cellulose-Schichten getrennt werden. Am besten gelingt die Trennung mit dem Laufmittel 2,5%ige Natriumcitratlösung : 25%iges Ammoniak (4 : 1). Die Abb. 3 und 4 zeigen Farbstofftrennungen auf Cellulose-Schichten in 30 Minuten.

Die Trennung von Anthrachinonfarbstoffen[6,7] gelingt auf Schichten aus acetyliertem Cellulosepulver. Wir verwenden die Sorten MN 300 Ac und MN 300 G/Ac. Die Entwicklungsdauer beträgt etwa 60 Minuten bei Anwendung des Laufmittels Äthylacetat : Tetrahydrofuran : Wasser (6 : 35 : 47). Tintenfarbstoffe[7] lassen sich auf Cellulose-Schichten MN 300 oder MN 300 G trennen mit der oberen Phase des Gemisches n-Butanol : Eisessig : Wasser (4 : 1 : 5).

Metallkationen trennte KLAMBERG[10] auf Cellulose-Schichten MN 300 und zwar nach Vortrennung in die klassischen Gruppen: Schwefelwasserstoffgruppe, Ammonsulfidgruppe und Erdalkaligruppe. Die für die Papierchromatographie ausgearbeiteten Trennvorschriften können weitgehend ohne Änderung übernommen werden. Die

Literatur S. 30

TABELLE 3
TRENNUNGEN VON METALLKATIONEN AUF CELLULOSE-SCHICHTEN

Schwefelwasserstoffgruppe
Schicht: Cellulosepulver MN 300
Laufmittel: n-Butanol mit $3\,N$ HCl gesättigt, Kammerübersättigung
Reihenfolge: Start — Ag Cu Pb Sb Bi As Cd Sn Hg — Front
Nachweisreagenzien: wässrige KJ-Lösung; H_2S–Wasser, Ammonsulfid-Lösung; Quercetin-Lösung + Ammoniakdampf

Ammonsulfidgruppe
Schicht: Cellulosepulver MN 300
Laufmittel: Eisessig : Pyridin : konz. HCl (80 : 6 : 20), Kammerübersättigung
Reihenfolge: Start — Cr Al Ni Mn Co Zn Fe — Front
Nachweisreagenzien: 1-(2-Pyridyl-azo)-2-naphthol in 0,2 %iger methanolischer Lösung + Ammoniakdampf; Oxin + Ammoniak: fluoreszierende Flecken im langwelligen UV-Licht

Erdalkaligruppe
Schicht: Cellulosepulver MN 300
Laufmittel: Methanol : konz. HCl : Wasser (8 : 1 : 1), Kammerübersättigung
Reihenfolge: Start — Ra Ba Sr Ca Mg Be — Front
Nachweisreagenzien: Oxin + Ammoniak + UV-Licht; Natriumrhodizonat; Strahlungsmessgerät für Ra

Flecken sind auch bei diesen Trennungen weit schärfer und kompakter als auf Papier. Die in Tabelle 3 angeführten Ionen der H_2S-Gruppe und der Erdalkaligruppe lassen sich in einem eindimensionalen Chromatogramm gleichzeitig trennen. Weitere Einzelheiten über Trennbedingungen sind in Tabelle 3 zusammengestellt.

H. GOEBELL[11] arbeitete eine Methode zur dc-Trennung von Tricarbonsäurezyklus-Substraten auf Cellulose-Schichten MN 300 aus. Die Schichten wurden durch Aufschlämmen des Cellulosepulvers in verdünnter Natronlauge auf pH 11,5 gebracht. Die Trennungen erfolgten zweidimensional und zwar in der 1. Richtung mit einem alkalischen Laufmittel, nämlich 95%iges Äthanol : konz. Ammoniaklösung : Wasser (8 : 2 : 1) und in der 2. Richtung mit einem sauren Laufmittel, nämlich der alkoholischen Phase des Gemisches Isobutanol : $5\,M$ Ameisensäure (2 : 3). Tabelle 4 gibt eine Zusammenstellung der R_F- und der R_{Malat}-Werte.

Zur quantitativen Auswertung werden die mittels Indikator (Bromkresolgrün) oder Radioaktivität lokalisierten Flecken ausgekratzt, in 0,5 bis 1 ml Wasser eluiert und im Eluat die Substratmengen quantitativ bestimmt. Die Bestimmung erfolgt im optisch-enzymatischen Test für Succinat, Fumarat, α-Ketoglutarat, Malat, dl-Isocitrat, Glutamat, Aspartat, Lactat, Pyruvat und chemisch, soweit möglich, für die übrigen Substrate. Die Radioaktivität wird mittels Autoradiographie lokalisiert und nach Auskratzen der Flecken in einem Flüssigkeits-Scintillationszähler (Tricarb) entweder direkt in dem Cellulosepulver oder indirekt in dessen Eluat bestimmt.

Über Zucker- und Uronsäuretrennungen auf Cellulose-Schichten MN 300 berichten A. SCHWEIGER UND R. GRAU[12,13]. Bei einer zweifachen Entwicklung mit Zwischen-

TABELLE 4

R_F- UND R_{Malat}-WERTE VON SUBSTRATEN DES TRICARBONSÄUREZYKLUS UND ANGRENZENDEN REAKTIONEN AUF SCHICHTEN AUS CELLULOSEPULVER MN 300

Säureanion	Alkalisches Laufmittel		Saures Laufmittel	
	$R_F \cdot 100$*	$R_{Malat} \cdot 100$*	$R_F \cdot 100$*	$R_{Malat} \cdot 100$*
Citrat	5	16	31	70
cis-Aconitat	7, 10	37, 54	29, 82	65, 185
dl-Isocitrat	8, 39	35, 198	28, 82	64, 184
Glyoxalat	14	65	70	142
Oxalacetat	15	75	20	42
Malonat	17	80	64	140
Aspartat	18	80	8	16
Malat	21	100	46	100
Glutamat	24	106	14	28
Succinat	27	123	80	174
α-Ketoglutarat	31	132	51	113
Fumarat	32	145	89	198
Lactat	56	213	80	172
Pyruvat	60	240	78	163
Fluoracetat	60	330	86	165
β-Hydroxybutyrat	66	320	88	194

* Durchschnittswerte von 10 Chromatogrammen.

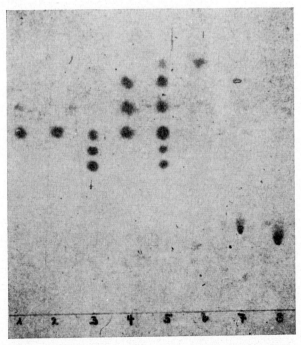

Abb. 5. Dünnschichtchromatogramm von Zuckern auf einer Schicht aus Cellulosepulver MN 300 (zweimalige Entwicklung mit Zwischentrocknen). 1 = Mannose; 2 = Arabinose; 3 = Galactose, Glucose, Mannose; 4 = Arabinose, Xylose, Ribose; 5 = Gemisch von 3, 4 und 6; 6 = Rhamnose; 7 = Glucuronsäure; 8 = Galacturonsäure.

Literatur S. 30

trocknen mit dem Laufmittel Äthylacetat : Pyridin : Wasser (2 : 1 : 2) wurden bei Zuckern die besten Trennungen erzielt (Vgl. Abb. 5). Der Nachweis erfolgt mit Anisidinphthalat; die Nachweisgrenze liegt bei etwa 0,3–0,5 γ. Uronsäuren konnten mit dem Laufmittel Isopropanol : Pyridin : Eisessig : Wasser (8 : 8 : 1 : 4) getrennt werden. Der Nachweis erfolgt ebenfalls mit Anisidinphthalat.

Ausser den in Abb. 5 gezeigten qualitativen Trennungen der Zucker Galactose, Glucose, Mannose, Arabinose, Xylose, Ribose und Rhamnose beschreiben R. GRAU UND A. SCHWEIGER[13] in einer Arbeit über den Nachweis von Quellstoffen in Fleischwaren wiederum Zuckertrennungen auf Cellulose-Schichten MN 300. Abb. 6 zeigt, dass halbquantitative Aussagen durch vergleichende Auswertung der Fleckengrösse bzw. der Fleckenfarbintensität möglich sind. Bei 1 sind aufgetragen 3 Zucker von je 0,6 γ, bei 2 je 1,25 γ und bei 3 je 2,5 γ. Bei 7 sind sogar nur Konzentrationen von je 0,3 γ aufgetragen und die Flecken sind noch klar erkennbar.

Die beiden Abb. 5 und 6 zeigen Chromatogramme von Modellgemischen von Zuckern. Abb. 7 zeigt dc-Trennungen von Quellstoff-Hydrolysaten auf Cellulose-Schichten MN 300. Man kann leicht erkennen, dass die hervorragenden Trennmöglichkeiten von Zuckern auf Cellulose-Schichten nicht nur bei Modellgemischen, sondern auch für praktische Untersuchungszwecke ausgenutzt werden können. Im einzelnen

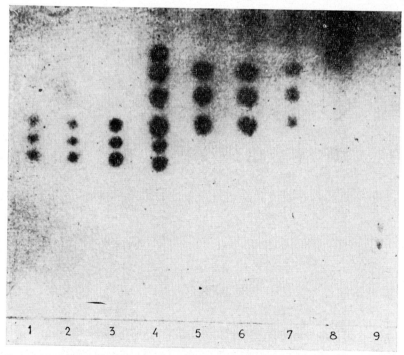

Abb. 6. Dünnschichtchromatogramm von Zuckern und Uronsäuren auf einer Schicht aus Cellulosepulver MN 300 (zweimalige Entwicklung mit Zwischentrocknen). D-Galactose, D-Glucose, D-Mannose: 1 = je 0,6 γ; 2 = je 1,25 γ; 3 = je 2,5 γ. L-Arabinose, D-Xylose, D-Ribose: 5 = je 1,25 γ; 6 = je 0,6 γ; 7 = je 0,3 γ; 8 = 2,5 γ Rhamnose; 9 = je 1,25 γ Galacturonsäure und Glucuronsäure; 4 = Gemisch aus 3, 5 und 8.

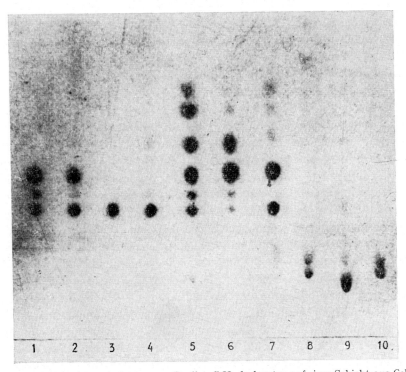

Abb. 7. Dünnschichtchromatogramm von Quellstoff-Hydrolysaten auf einer Schicht aus Cellulosepulver MN 300 (zweimalige Entwicklung mit Zwischentrocknen). 1 = Johannisbrotkernmehl (Carubin); 2 = ,,Nurugum" (Guaran); 3 = Agar; 4 = Carrageen; 5 = Testgemisch (von unten Galacturonsäure; Galactose, Glucose, Mannose und Arabinose nicht getrennt, Xylose, Ribose, Rhamnose); 6 = Traganth; 7 = Gummi arabicum; 8 = Na-Alginat; 9 = Pektin; 10 = Galacturonsäure und Glucuronsäure.

handelt es sich um Hydrolysate von Johannisbrotkernmehl, Nurugum, Agar, Carageen, Traganth, Gummi arabicum, Natriumalginat und Pektin.

In Ergänzung zu den dc-Zuckertrennungen auf Cellulose-Schichten ist noch anzuführen, dass die Fructose bei dem genannten Laufmittel Äthylacetat : Pyridin : Wasser (2 : 1 : 2) den gleichen R_F-Wert aufweist wie die Arabinose und deshalb eine Auftrennung von Fructose und Arabinose bei den erwähnten Bedingungen nicht möglich ist. Liegen Fructose und Arabinose nicht gleichzeitig vor, so ist eine Unterscheidung durch die unterschiedliche Anfärbung möglich[14].

W. MOHR[15] versuchte qualitativ monomere Polyhydroxyphenole aus Schokolade-Extrakten papierchromatographisch oder dünnschichtchromatographisch auf Kieselgel-Schichten zu trennen. Beide Methoden führten jedoch trotz Vorreinigung der Extrakte an Polyamidsäulen nicht zum Erfolg. Erst die Chromatographie auf Cellulose-Schichten MN 300 ermöglichten eine einwandfreie Trennung von (+)-Catechin, (—)-Epicatechin und Pro-anthocyanidin L_1. Als Laufmittel benutzte er die obere Phase des Gemisches n-Butanol : Eisessig : Wasser (4 : 1 : 5). Die Anfärbung erfolgte mittels Phosphormolybdänsäure.

Da die qualitativen Trennversuche zu sehr günstigen Trenneffekten führten, unter-

Literatur S. 30

suchte der genannte Autor, ob auf dieser Basis eine halbquantitative Methode durch vergleichende Auswertung der Fleckengrösse möglich wäre. Er kommt zu dem Ergebnis, dass dies möglich ist, wobei die maximalen Abweichungen bei $\pm 7\%$ liegen.

Weiterhin berichtet W. Mohr über eine quantitative Bestimmung der genannten Polyhydroxyphenole und zwar muss man dazu wie folgt verfahren: die getrennten Phenole werden mit diazotierter Sulfanilsäure angefärbt, die Flecken ausgekratzt und das Pulver wird mit Äthanol eluiert. Von der so gewonnenen gelbgefärbten Lösung wird spektralphotometrisch die Extinktion gemessen.

Trennungen von Nucleinsäurederivaten auf Schichten auf Cellulosebasis wurden von K. Randerath[16-20] in einer Anzahl Arbeiten beschrieben. Die Anwendung der DC bringt gegenüber den bisherigen Trennmethoden, nämlich der Papierchromatographie und Hochspannungselektrophorese, eine beträchtliche Herabsetzung der Analysendauer und eine Steigerung der Nachweisempfindlichkeit. Bei anorganischen Trennschichten ergeben sich bei so extrem hydrophilen Verbindungen wie Nucleosidphosphate Schwierigkeiten, die sich auch durch die Anwendung stark polarer Laufmittel nicht beseitigen lassen. Für die DC der Nucleotide, Nucleobasen und Nucleoside sind Schichten auf Cellulosebasis, d.h. sowohl normale Cellulose als auch Cellulose-Anionenaustauscher, wesentlich besser geeignet. Auf diesen Schichten können verschwindende Spuren dieser Verbindungen unmittelbar im UV-Licht erkannt werden.

Ein Beispiel für die Überlegenheit der Chromatographie von Nucleobasen und Nucleosiden[19] auf Cellulose-Schichten MN 300 G gegenüber Chromatographiepapier zeigt Abb. 8. Abb. 8a zeigt eine Trennung auf der Cellulose-Schicht, Abb. 8b einen Trennversuch bei völlig gleichen Bedingungen auf Papier. Als Laufmittel dient destilliertes Wasser.

Nucleotide trennte Randerath auf Schichten aus normalem Cellulosepulver MN

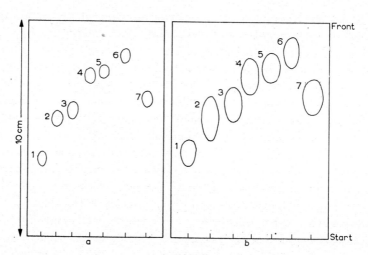

Abb. 8. (a) Cellulose-Schichtchromatogramm von Nucleobasen und Nucleosiden. Laufmittel: dest. Wasser. Laufstrecke: 10 cm in 45 Min. (b) Unter identischen Bedingungen hergestelltes Papierchromatogramm. Laufstrecke: 10 cm in 46 min. Papier: Ederol 202. 1 = Adenin; 2 = Adenosin; 3 = Hypoxanthin; 4 = Inosin; 5 = Uracil; 6 = Uridin; 7 = 6-Chlorpurin.

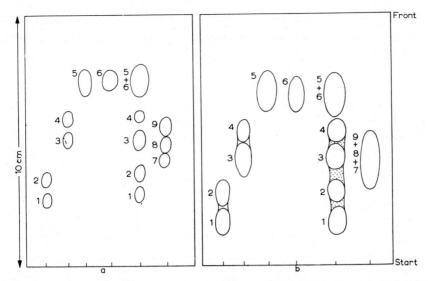

Abb. 9. (a) Cellulose-Schichtchromatogramm von Nucleotiden. Laufstrecke: 10 cm in 91 Min. (b) Papierchromatogramm, unter identischen Bedingungen hergestellt. Laufstrecke: 10 cm in 134 Min. Papier: Schleicher & Schüll 2043b. 1 = 3'-AMP; 2 = 2'-AMP; 3 = 3'-GMP; 4 = 2'-GMP; 5 = 2'- + 3'-CMP; 6 = 2'- + 3'-UMP; 7 = 5'-AMP; 8 = ADP; 9 = ATP.

300[18] und MN 300 G[19,21] sowie auf Schichten aus den Anionenaustauschern DEAE[21] und ECTEOLA[16,17,21]. Abb. 9 veranschaulicht ähnlich wie Abb. 8 den Unterschied in der Trennschärfe zwischen einer Schicht aus Cellulosepulver (Abb. 9a) und Papier (Abb. 9b). Die eindeutig bessere Trennung ergibt die Schicht. Die Trennbedingungen wurden wiederum völlig gleich gehalten. Als Laufmittel diente ein Gemisch aus gesättigter Ammonsulfatlösung, 1 M Natriumacetatlösung und Isopropanol (80 : 18 : 2). Die Entwicklungsdauer betrug 90 Minuten.

Die Trennungen von Nucleotiden auf Austauscher-Schichten sind auf Grund ihrer ionenaustauschenden Wirkung naturgemäss noch schärfer und schneller als auf normalen Cellulose-Schichten. Als Laufmittel für DEAE- und ECTEOLA-Schichten kann man verwenden: Salzsäuren verschiedener Konzentration, etwa in der Grössenordnung von 0,01 bis 0,1 N, oder aber verdünnte Kochsalzlösungen ähnlicher Konzentration.

Die quantitative Bestimmung der Nucleinsäurederivate erfolgt nach Ablösen der Substanzflecken und anschliessender Elution UV-spektroskopisch. Als Elutionsmittel wählt man solche Pufferlösungen, in denen die Nucleotide besonders leicht löslich sind. Im übrigen verfährt man nach der in der Papierchromatographie üblichen Technik.

Polycyclische Kohlenwasserstoffe konnten auf Schichten aus acetyliertem Cellulosepulver von verschiedenen Autoren[22,23] getrennt werden. BADGER und Mitarbeiter verwenden als Laufmittel (bei Kammersättigung) Methanol : Äther : Wasser (4 : 4 : 1) oder Toluol : Äthanol : Wasser (4 : 17 : 1), wovon jedoch das erstere Laufmittel dem zweiteren vorzuziehen ist wegen der kompakteren Flecken. Der Nachweis erfolgt mittels UV-Licht.

Literatur S. 30

Abschliessend noch einige generelle Anmerkungen zur DC auf Cellulose-Schichten: Schichten aus normaler Cellulose bzw. Cellulosederivaten haben sich bereits gut bewährt bei der Trennung stark hydrophiler als auch bei hydrophoben Verbindungen. Die Nachweisempfindlichkeit gegenüber Chromatographiepapier ist im allgemeinen um eine bis zwei Zehnerpotenzen höher; die Trennzeiten sind erheblich verkürzt. Im Vergleich zu den anorganischen Schichten ist die Haftfestigkeit von Cellulose-Schichten ausserordentlich hoch. Generell kann empfohlen werden, bei der Wahl des Trennmediums stets beweglich zu bleiben, weil gerade bei den zahlreichen Substanzen, die sich nicht streng in die Gruppen hydrophil oder hydrophob einordnen lassen, die Trennungen mal auf anorganischen mal auf organischen Schichten besser verlaufen.

LITERATUR

1. K. TEICHERT, E. MUTSCHLER UND H. ROCHELMEYER, Deut. Apotheker-Ztg., 100 (1960) 283.
2. K. TEICHERT, E. MUTSCHLER UND H. ROCHELMEYER, Deut. Apotheker-Ztg., 100 (1960) 477.
3. K. TEICHERT, E. MUTSCHLER UND H. ROCHELMEYER, Z. Anal. Chem., 181 (1961) 325.
4. P. WOLLENWEBER, J. Chromatog., 9 (1962) 369.
5. E. VON ARX UND R. NEHER, J. Chromatog., im Druck (siehe Nachtrag).
6. P. WOLLENWEBER, J. Chromatog., 7 (1962) 557.
7. E. STAHL, Dünnschicht-Chromatographie, Ein Laboratoriumshandbuch, Springer Verlag, Berlin, 1962, Seite 359.
8. P. WOLLENWEBER, Mitt. Blatt GDCh-Fachgr. Lebensmittelchem. u. Gerichtl. Chem., 17 (1963) 65.
9. K. RANDERATH, Dünnschicht-Chromatographie, Monographie zu ,,Angewandte Chemie" und ,,Chemie-Ingenieur-Technik", Nr. 78, Verlag Chemie GmbH, Weinheim/Bergstr., 1962, Seite 203.
10. H. KLAMBERG, vgl. 9, jedoch Seite 221.
11. H. GOEBELL UND M. KLINGENBERG, Chromatographie Symposium II, Société Belge des Sciences Pharmaceutiques, Bruxelles, 1962, Seite 153.
12. A. SCHWEIGER, J. Chromatog., 9 (1962) 374.
13. R. GRAU UND A. SCHWEIGER, Z. Lebensm. Untersuch. Forsch., 119 (1963) 210.
14. A. SCHWEIGER, Privatmitteilung.
15. W. MOHR, Fette, Seifen, Anstrichmittel, 64 (1962) 739, 831.
16. K. RANDERATH, Angew. Chem., 73 (1961) 436.
17. K. RANDERATH, Angew. Chem., 73 (1961) 674.
18. K. RANDERATH UND H. STRUCK, J. Chromatog., 6 (1961) 365.
19. K. RANDERATH, Biochem. Biophys. Res. Commun., 6 (1961/62) 452.
20. K. RANDERATH, Nature, 194 (1962) 768.
21. K. RANDERATH, Angew. Chem., 74 (1962) 484; Angew. Chem. Intern. Ed. Engl., 1 (1962) 435.
22. T. WIELAND, G. LÜBEN UND H. DETERMAN, Experientia, 18 (1962) 430.
23. G. M. BADGER, J. K. DONNELLY UND T. M. SPOTSWOOD, J. Chromatog., 10 (1963) 397.

NACHTRAG

Während der Drucklegung sind noch eine Anzahl Publikationen über DC-Trennungen auf Cellulose-Schichten erschienen, die nachstehend zunächst stichwortartig, dann in Form von ausführlichen Literaturzitaten aufgeführt werden:

Getrennte Substanzen	Literatur
Acetessigsäure/Aceton	33
Acylneuraminsäuren	40
Aminosäuren	5
Anthocyanine	27

Ganglioside 32
2-Hydroxybenzophenone und Derivate 35
Kationen, anorg. 30, 31
Nucleinsäure-Derivate 24, 25, 26, 36, 37, 38, 39
Phosphate, kondensierte 29
Zucker 28, 34

LITERATUR

5. E. VON ARX UND R. NEHER, *J. Chromatog.*, 12 (1963) 329.
   Eine multidimensionale Technik zur chromatographischen Identifizierung von Aminosäuren.
24. G. N. MAHAPATRA UND O. M. FRIEDMAN, *J. Chromatog.*, 11 (1963) 265.
    Separation of isomeric methylated deoxyguanosines on thin cellulose layers prepared with a glass rod applicator.
25. R. G. COFFEY UND R. W. NEWBURGH, *J. Chromatog.*, 11 (1963) 376.
    The effect of calcium sulfate as the binder in DEAE-cellulose thin-layer chromatography for separating nucleic degradation products.
26. T. A. DYER, *J. Chromatog.*, 11 (1963) 414.
    The separation of the nucleotides of an alkaline hydrolysate of ribonucleic acid by thin-layer chromatography.
27. N. NYBOM, *Fruchtsaft-Ind.*, 8 (1963) 205.
    Dünnschichtchromatographische Anthocyanin-Analyse von Fruchtsäften.
28. E. V. DJATLOVICKAJA, V. V. VORONKOVA UND D. D. BERGEL'SON, *Dokl. Akad. Nauk SSSR*, 145 (1962) 325.
    Dünnschichtchromatographische Trennung von verschiedenen Polyalkoholen und Monosacchariden auf Cellulose-Schichten.
29. T. RÖSSEL, *Z. Anal. Chem.*, 197 (1963) 333.
    Die chromatographische Analyse von Phosphaten.
    Teil 2: Die Dünnschichtchromatographie der kondensierten Phosphate.
30. H. HAMMERSCHMIDT UND M. MÜLLER, *Papier*, 17 (1963) 448.
    Ein dünnschichtchromatographisches Verfahren zur Identifizierung von Papierfüllstoffen und Streichmassen.
31. F. W. H. M. MERKUS, *Pharm. Weekblad*, 98 (1963) 947.
    The use of thin-layer chromatography in toxicological analysis of metals.
32. E. KLENK UND W. GIELEN, *Z. Physiol. Chem.*, 330 (1963) 218.
    Über ein zweites hexosaminhaltiges Gangliosid aus Menschengehirn.
33. M. RINK UND S. HERRMANN, *J. Chromatog.*, 12 (1963) 249.
    Nachweis von Acetessigsäure und Aceton (nach Überführung in ihre 2,4-DNP-Hydrazone) im Harn mit Hilfe der Dünnschichtchromatographie.
34. L. D. BERGEL'SON, E. V. DJATLOVICKAJA UND V. V. VORONKOVA, *Dokl. Akad. Nauk SSSR*, 149 (1963) 1319.
    Descending thin-layer chromatography of polyhydroxy compounds on cellulose layers.
35. E. KNAPPE, D. PETERI UND I. ROHDEWALD, *Z. Anal. Chem.*, 197 (1963) 364.
    Imprägnierung chromatographischer Dünnschichten mit Polyestern: Trennung und Identifizierung substituierter 2-Hydroxybenzophenone und andere U.V.-Absorber.
36. K. RANDERATH, *Angew. Chem.*, 74 (1962) 780.
    Ein einfaches Herstellungsverfahren für Cellulose-Anionenaustauscher und Anionenaustauscherpapiere.
37. K. RANDERATH, *Dünnschicht-Chromatographie*, Verlag Chemie, Weinheim, 1962, Seite 35 bzw. 193.
    Dünnschichtchromatographische Trennung von Nucleotiden an Polyäthylenimin-(PEI) Cellulose.
38. K. RANDERATH, *Biochim. Biophys. Acta*, 61 (1962) 852.
    PEI-Cellulose — ein neuer Anionen-Austauscher für die Chromatographie von Nucleotiden.
39. K. RANDERATH UND E. RANDERATH, *J. Chromatog.*, 10 (1963) 509.
    Polyphosphat-Cellulose — ein neuer Kationenaustauscher. Herstellung und Anwendung in der Dünnschichtchromatographie. Trennung von Nucleobasen und Nucleosiden.
40. H. FAILLARD UND J. A. CABEZAS, *Z. Physiol. Chem.*, 333 (1963) 266.
    Isolierung von N-Acetyl- und N-Glykolylneuraminsäure aus Kälber- und Hühnerserum.

THIN-LAYER CHROMATOGRAPHY ON LOOSE LAYERS OF ALUMINA

LUDVÍK LÁBLER

Institute of Organic Chemistry and Biochemistry, Czechoslovak Academy of Science, Praha (Czechoslovakia)

1. INTRODUCTION

Alumina is used as an adsorbent mainly in column chromatography but it is not used to the same extent as silica gel for thin-layer chromatography. The number of papers reporting the application of alumina in thin-layer chromatography are far lower than those for silica gel.

Alumina is used for thin-layer chromatography in the form of a powder which may or may not contain a binder. The binding agent is usually gypsum. A suitable adsorbent containing approximately 5% gypsum is available commercially from E. Merck of Germany and Fluka of Switzerland. Recently, alumina which does not contain any binder appeared on the market and is supplied by Woelm of Eschwege, Germany, and by Serva-Entwicklungslabor of Heidelberg, Germany. An adsorbent which was not originally intended for thin-layer chromatography was used by PEIFER[1] who chose a particle size of 200 mesh and mixed it with 5% of gypsum. HUNECK[2] recommends the so-called fibre alumina for preparing alumina layers without a binder; according to the author, the layers should have improved mechanical properties. The slurry, *i.e.* a suspension of alumina in water, is prepared in a similar manner to that of silica gel using about 1 part of alumina and 2–3 parts of water. This proportion varies slightly depending on the brand used. The thin layers are prepared in the same way as those of silica gel by spreading the slurry with one of the known applicators. Organic solvents may be used instead of water for preparing the slurry; this technique was used for covering microscope slides by dipping them into the suspension. The plates covered with alumina are allowed to dry in the air and then activated at an elevated temperature. The activated plates have to be stored in a desiccator over calcium chloride etc. If an adsorbent of a particular activity is required it may be obtained by either altering the activation process or by placing plates of maximum activity for some time in a desiccator containing alumina of the desired activity determined according to BROCKMANN AND SCHODDER[3]. Solutions of samples are applied to the plates in the usual manner and the technique for developing the chromatograms does not differ from that used for silica-gel chromatography.

Alumina also finds its use in the so-called loose layers, a modification of thin-layer chromatography. Here, unlike thin-layer chromatography, the adsorbent is not spread in the form of a suspension, which after drying adheres to the glass surface, but is spread as a dry powder. This method has not extended to such a degree as thin-layer chromatography however, but good results have been obtained and numerous papers published. This technique will also be mentioned therefore in this paper. The method has been used for a relatively long time. It was first described by the Swiss chemists MOTTIER AND POTTERAT[4] who used circular development and later linear development in the form of that used in chromatography today. The method escaped notice however and was used six years later by the Russian chemist MISTRYUKOV[5].

The limits of this method are now defined more precisely below. The adsorbent used for preparing loose layers has larger particles than that used for thin-layer chromatography. It is immediately clear that the coarsest adsorbent will not give as sharp separations as the powdery material used in normal thin-layer chromatography. Chromatography on loose layers is therefore not well suited for separating highly complicated mixtures of compounds which have similar R_F values. For normal work, however, the technique is fully convenient and gives satisfactory results. It has the advantage that the layers are prepared with very simple equipment which can be made easily in the laboratory. As the chromatograms are not stored prior to use but prepared directly before the experiment, an adsorbent of any activity can be used. The alumina has the same particle size as that used for column chromatography and therefore the material for preparing loose layers is readily available in any laboratory where column chromatography is performed. Loose layers can also be prepared promptly from any other adsorbent. The alumina used for both column and thin-layer chromatography is of the same particle size and therefore the results from the latter can be applied to the former. The main area of loose layers however, is in preparative work and this will be mentioned later.

2. TECHNIQUE

The technique of loose layers will now be discussed in more detail. The glass plates are usually of the same size as for thin-layer chromatography, *i.e.* 20 cm long and 5, 10 or 20 cm wide. The size of the plates is by no means restricted however and can be quite arbitrary. The adsorbent, in this case alumina, has a particle size of approximately 75–100 μ[6,7,8]. The size was determined empirically. Finer material cannot be used because on decreasing the particle size the pouring properties of the adsorbent change and the layers cannot be prepared. An applicator which can be prepared in the laboratory (Fig. 1) is used for covering the plates. The device consists of a glass rod or thick walled tube approximately 1.2 cm in diameter which is fitted with two rings made of rubber or polythene tubing. The distance between the inner edges of the rings has to be 1 cm shorter than the width of the plate. The thickness of the rings defines the thickness of the layer[6]. The rings can also be made from adhesive tape or Hansaplast[7], the number of tiers determining the thickness of the layer. A similar apparatus can be turned out from metal. The most convenient thickness of the layer is 0.6 mm. The layer is prepared by spreading the adsorbent uniformly to

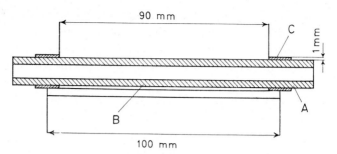

Fig. 1. Device for preparation of adsorbent layer.

a depth of 3–4 mm on the horizontal plate. The rod is then moved from one edge of the plate to the other (Fig. 2). The excess adsorbent is removed and a uniform thin layer results with margins of about 5 mm on both sides. Another method is to attach glass strips to the longer sides of the plate thus forming a shallow trough (Fig. 3). The adsorbent is poured between the strips and the layer is produced by drawing a glass rod or ruler, resting on the glass strips, across the plate.

Obviously the chromatograms must be handled differently from normal thin-layer plates. It should be mentioned, however, that a loose layer is far less sensitive to disturbance caused by tilting the plates than might be thought at first sight. The layer does not start to slide until the plate is inclined at an angle of 40° and this is never encountered in practice. The layer is very sensitive to shaking and vibrations must be carefully avoided.

The solutions of samples are applied with a micropipette in the usual manner and volatile solvents quickly evaporate.

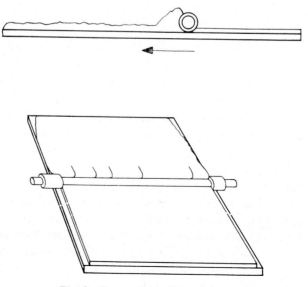

Fig. 2. Preparation of loose layers.

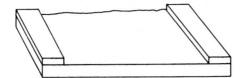

Fig. 3. Trough from glass.

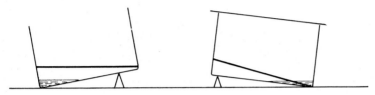

Fig. 4. Glass tank.

Fig. 5. Another position of plate and tank.

Fig. 6. Overflow technique using a mound of adsorbent.

The chromatograms are developed in glass tanks with the plates slightly inclined at an angle of approximately 20° (Fig. 4). The plates lean either directly against the wall of the tank or are supported in a suitable way (Fig. 5). The solvent is at the bottom of the tank and the plates are developed using the ascending technique. As the adsorbent is coarse grained the development takes a relatively short time, information about the course of a reaction may therefore be rapidly obtained. The efficiency of the separation determines the length of the chromatogram which in most cases is 15–17 cm.

An overflow technique has been developed for loose layers which is similar to that used for bound layers. However the method of BRENNER AND NIEDERWIESER[9] may be used in principle. I would like to mention some other possibilities here. The simplest procedure (Fig. 6) uses the so-called mound of adsorbent into which the

References p. 43

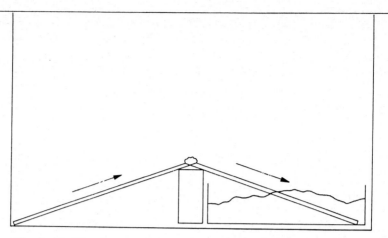

Fig. 7. Modification of the technique (see Fig. 6) suggested by Schwarz.

excess solvent is sucked[10]. As the capacity of this device is relatively low, another modification based on the same principle, was suggested by SCHWARZ[10] (Fig. 7). Two plates covered with adsorbent, one ascending the other descending, are used, with passage from one to the other being facilitated by a small mound. The excess solvent is sucked into adsorbent in a Petri dish into which the second plate is inserted. A third device was recently reported by MISTRYUKOV[11] (Fig. 8) but this is relatively complicated. The development is descending, the plate being supported by a frame made from aluminium. The transfer of the solvent is facilitated by strips of filter paper inserted into the upper and lower slit. Alternatively, the distance of the slits is carefully regulated to ensure the rise of the solvent by capillary force up to the layer and down by gravity into the tank. Both systems according to the author are equally satisfactory. The solvent reaches the layer, descends through it and finally is collected on the bottom of the tank.

The visualisation of the spots is carried out in principle in the same manner as with normal thin-layer plates. The layer may be exposed both to the vapours of a reagent *e.g.* iodine, or sprayed with a reagent. We have only to bear in mind that the layer is not fixed and that the process requires some special conditions. The plate for

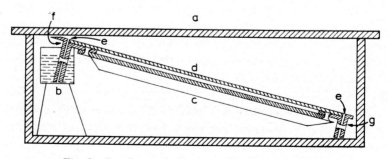

Fig. 8. Overflow technique according to Mistryukov.

detection is nearly horizontal, and the spraying is carried out from a distance of about 50 to 75 cm in such a way that the finely dispersed reagent falls down on the plate by its own weight. If possible, the detection is carried out with layers which are still moist with the solvent. In order to make detection easier, with corrosive reagents in particular, MISTRYUKOV[12] suggested a device for spraying chromatograms. It consists of a glass cover into which the plate is inserted; the sprayer is located on the side and its outlet is turned towards the dome of the cover. The dispersed reagent falls down by its own weight and there is no undesirable air turbulence.

A very advantageous method of detection is to use adsorbent impregnated with a fluorescent substance. In the case of alumina morin is used. This is a well known analytical reagent which gives stable fluorescent compounds with aluminium derivatives. Many years ago BROCKMANN AND VOLPERS[13] used alumina impregnated with morin for column chromatography. The adsorbent is prepared by mixing alumina (500 g) with a methanolic solution of morin (150 mg). The yellow morin solution becomes colourless and the adsorbent is collected by Buchner filtration, dried, and activated at 110°. The adsorbent is yellowish and gives a strong fluorescence in ultraviolet light. The reagent is very firmly bound and cannot be eluted even by alcohol; this is of special importance for preparative work. Compounds such as α,β-unsaturated ketones, which adsorb in the 240 mμ range, appear in the light of short wave sources, e.g. chromatolite*, as dark spots on a bright yellow fluorescent background[6]. It has been found on the other hand that steroids, compounds which do not adsorb ultraviolet light, can be detected on morin-impregnated alumina layers as well[6]. The developed chromatograms must be perfectly dried under an infra-red lamp and viewed in the light of Philora. The compounds appear as bright yellow spots on a darker background. In no case in the steroid field did this means of detection fail. Detection may also be carried out by spraying normal alumina layers with a very dilute methanolic morin solution (0.005–0.01 %)[6].

It has already been mentioned that the technique of loose layers allows one to use alumina of any activity which beforehand was determined according to BROCKMANN[3]. The activity may also be determined directly according to HEŘMÁNEK, SCHWARZ AND ČEKAN[14] by thin-layer chromatography. The authors used a mixture of azo dyes: p-methoxyazobenzene, sudan yellow, sudan red, and p-aminoazobenzene, and after developing the chromatoplate with carbon tetrachloride compared the R_F values with those obtained on alumina of known activities determined according to BROCKMANN[3]. The results are summarised in Table 1 according to which the activity of alumina can be easily determined from the R_F values of the dyes.

Alumina used for loose layers has the same particle size as the adsorbent used for column chromatography. It follows that by means of thin-layer chromatography we can obtain exact information about the behaviour of a mixture of compounds on a column. One important thing must not be forgotten when we apply the results from thin-layer chromatography to column chromatography. In thin-layer chromatography

---

* Hanovia, Slough, Bucks., England.

References p. 43

TABLE 1

R_F VALUES OF AZO DYES ON ALUMINA

(according to HEŘMÁNEK, SCHWARZ AND ČEKAN[14])

Azo dye	Activity according to BROCKMANN AND SCHODDER			
	II	III	IV	V
azobenzene	0.59	0.74	0.85	0.95
p-methoxyazobenzene	0.16	0.49	0.69	0.89
sudan yellow	0.01	0.25	0.57	0.78
sudan red	0.00	0.10	0.33	0.56
p-aminoazobenzene	0.00	0.03	0.08	0.19

a large excess of adsorbent is used with respect to the sample. This means that the same solvent system cannot be used for the column, unless the proportion of sample to adsorbent amounts to at least 1 : 500. If the usual proportion of 60–100 parts of adsorbent is used, all compounds with R_F higher than 0.1 will be eluted together in the first fractions. It is therefore necessary to start with a solvent less polar than that used for thin-layer chromatography and to add the more polar solvent carefully in small portions.

The area in which chromatography on loose layers finds its use and where good results are achieved is the preparative technique. Generally, it may be said that preparative separation of mixtures up to 0.5–1 g is more convenient and faster on thin layers than on columns. For preparative work, the same sort of alumina is used as that for analytical chromatograms, only the thickness of the layer is increased to 1–2 mm. The layers are prepared analogously. We do not prefer thicker layers because there is always a danger of irregular development of the solvent through the latter. This deforms the shapes of the zones and also the application of samples is rather difficult in order to let the mixture adsorb uniformly in the whole cross-section of the layer. The size of the plates are usually 20 × 20 cm; also far larger plates are used, up to 50 cm in length and of great capacity[15], by means of which outstanding results are obtained. The handling of the plates, however, has some drawbacks. First of all certain experience is needed to manipulate the plates, second, big chromatographic tanks have to be used, and third, manipulating the chromatograms which contain a considerable amount of mostly inflammable solvents after development, is less convenient. Although the big plates give very good results, it is still advisable to use a corresponding number of smaller plates. The solutions of samples are applied on the starting line in the form of separate spots of uniform concentration. An application in the form of spots so close together that they form a line is not recommended because uniform concentration cannot be guaranteed along the line and irregularities in the shapes of the zones result. The chromatograms are developed in the usual manner. As for detection, the most suitable method in which no material is lost, is to use morin-impregnated alumina. Otherwise, a reagent may be sprayed on the chromatogram, partly covered with a sheet of stainless steel. Even iodine vapours are

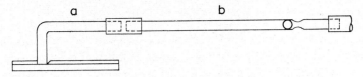

Fig. 9. Device for collecting adsorbent zones.

recommended as a harmless reagent because the detected zones in many cases can be eluted without any chemical change. It is also possible to insert a narrow strip of filter paper into the still moist layer and after allowing the solvent to rise a little, to carry out the detection on the paper. The zones are collected by means of a simple device (Fig. 9) which consists of two glass tubes connected by rubber tubing. Into the contracted part, a cotton plug is inserted. The device is connected to a water-pump and the zones are collected as with a vacuum cleaner. For larger amounts, an apparatus is used which is connected to a flask (Fig. 10). The collected adsorbent is put into a chromatographic tube and eluted with a polar solvent to constant weight of the eluate. In any case it is necessary to check the purity of any isolated component by analytical thin-layer chromatography.

To prevent loss of material it is advisable to collect and to elute a narrow band of adsorbent before and after the zone. The capacity of a plate varies, depending upon the type of separation and the differences in R_F values. The optimum concentration per spot must be determined beforehand in the same way as in paper chromatography.

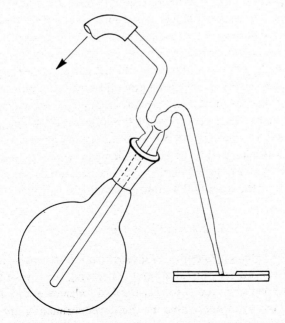

Fig. 10. Device for collecting adsorbent zones (larger amounts).

References p. 43

3. APPLICATIONS

It is not possible or even reasonable to present here a detailed survey of all the cases where alumina has been used to effect separations in different fields of organic chemistry. I shall therefore give a few examples which are of a general informative character.

Much work has been done in the field of alkaloids. From the chromatographic point of view the fundamental work was done by WALDI et al.[16]. They investigated several groups of alkaloids and used alumina as well as other adsorbents. Because alumina shows a tendency to cause tailing of the spots, WALDI added a strong organic base, e.g. diethylamine (0.05%) to the solvent system. The base deactivated the adsorbent enabling WALDI to use less polar solvent systems.

Among others, opium[17,18], belladona and opium[19] alkaloids were investigated and also the bases from *Vinca minor*[20], *Catharantus roseus*[21], *Withaurea somnifera*[22], *Lobelia*[23], *Voacanga africana*[24], and the Mexican drug Ololiuqui.

A number of papers have been published on the chromatography of steroids.

In the vitamin field the following mixtures of fat-soluble compounds were separated on alumina layers: tocopherols[26]; vitamin A and D_2; vitamin A and β-carotene; vitamin A acetate, β-carotene, α-tocopherol acetate and vitamin D_2[27]. The antibiotics epi-griseofulvin and racemic griseofulvin were separated using chloroform as the mobile phase[26]. Alumina was used in analysing pharmaceuticals, e.g. analgetics[17] and barbiturates[28]. The latter compounds were separated on the so-called Alusil which is a mixture of equal parts of alumina and silica gel[28]. Alusil was also used for chromatographing indicators[29] and sugars[30]. Polyphenyls[31], chlorinated hydrocarbons possessing insecticidal activity[32], isomeric cresols[33], nitrophenols[8,34] and nitroanilines[8] were also separated on alumina layers. Very good results were obtained in the field of terpene chemistry both with monoterpenes and sesquiterpenes. In the latter group, hydrocarbons of the eremophillane type were separated[35]. The biosynthesis of labelled carotol was studied[36] and some derivatives of the germacrane type were chromatographed[37]. Preparative thin-layer chromatography was the only possible means of isolating the unstable sesquiterpenic alcohol cryptoacoronol which isomerises on a column because of the long contact with the adsorbent[38]. Good results were also obtained by HUNECK[2] who separated a series of triterpenic compounds on alumina layers. Alumina was also used for separating stereoisomeric Cr- and Co-complexes of azodyes[39] and fat-soluble dyes[40]. Volatile carbonyl compounds were chromatographed on alumina layers in the form of 2,4-dinitrophenylhydrazones[41].

4. INFLUENCE OF THE SOLVENT

I would like to take this opportunity to present here, in a concise form, some previously unpublished results of general interest concerning the role of hydrogen bridges which are formed between the chromatographed substance and the solvent. This study was carried out by means of alumina thin-layer chromatography by HEŘMÁNEK, SCHWARZ AND ČEKAN[34] from the Institute for Natural Drugs in Prague.

TABLE 2
SOLVENT SEQUENCES FOR ALUMINA

Trappe	Strain	Jacques, Mathieu	Reichstein	Neher
light petroleum	light petroleum	hexane	pentane	pentane
cyclohexane	cyclohexane	benzene	light petroleum	light petroleum
carbon tetrachloride	carbon disulphide	ether	benzene	carbon tetrachloride
trichloroethylene	ether	chloroform	ether	benzene
toluene	acetone	ethyl acetate	chloroform	ether
benzene	benzene	1,2-dichloroethane	methanol	methylene chloride
methylene chloride	toluene	2-butanol		chloroform
chloroform	ethyl acetate	acetone		acetone
ether	1,2-dichloroethane	ethanol		ethyl acetate
ethyl acetate	alcohols	methanol		n-propanol
acetone	water			ethanol
n-propanol	pyridine			methanol
ethanol	org. acids			water
methanol				acetic acid

For the quality of separation of compounds on alumina, the type of eluent is of decisive importance. The choice is especially important in column chromatography because the compounds have to be eluted by a series of solvents of increasing elution power. For this purpose, several authors proposed an eluotropic series (Table 2) and applied different criteria for evaluating the solvents. TRAPPE[42] studied the stationary equilibrium between adsorbent and the solution for some less polar compounds and for different solvents; STRAIN[43] carried out a similar comparison but with polar compounds. JACQUES AND MATHIEU[44] found that the adsorbtivity of a solvent is proportional to its dielectric constant ε. REICHSTEIN and co-workers[45,46] and NEHER[47] arranged the sequence of solvents empirically.

When the sequences are compared, one sees that the same solvents are placed in different orders. According to TRAPPE, ether is a more powerful eluent than chloroform, however, JACQUES AND MATHIEU, and REICHSTEIN AND NEHER, place ether before chloroform. The series devised by STRAIN shows still greater differences, thus ether and acetone are placed before benzene. These discrepancies were in some cases assumed to be due to errors but in principle have not yet been elucidated.

No explanation for these differences has been given, mainly because up to the present time little attention has been paid to the related influence of adsorbent, substance and solvent. Without this consideration a uniform and generally valid sequence of solvents cannot be arranged.

The authors of this study observed that the adsorbtivity of a series of compounds was changed by using different solvents. They found that for some compounds, chloroform was a more powerful eluent than ether but for other substances the

References p. 43

reverse was true. From this example it was quite obvious that dielectric constants cannot be a sufficient criterion for elution capacity, as thought by JACQUES AND MATHIEU, especially as the constants for both solvents do not differ very much (ether 4.47, chloroform 5.2). It was clear that some other properties were responsible for the difference in the elution capacity.

HEŘMÁNEK et al. studied the influence of solvent on the behaviour of compounds chromatographed on alumina. They assumed that the speed of migration is affected by a number of factors, e.g. van der Waals forces, hydrogen bridges, mutual influence of dipoles, coordination, salt formation etc., by which the compound is bound both to the adsorbent and the solvent.

One of these factors is hydrogen bridges. From this point of view, the more firmly bound the compound by a hydrogen bridge to the solvent, the faster the compound is eluted. HECKER[48] divides organic compounds into five fundamental groups according to the manner in which the compounds contribute to the formation of a hydrogen bridge. Compounds which contain an acidic hydrogen in the molecule and form a hydrogen bridge with bases are called hydrogen donors (e.g. chloroform, methylene dichloride, 1,2-dichloroethane). Compounds with a free electron pair, i.e. bases, are called hydrogen acceptors (e.g. ethers, ketones, nitro compounds, esters, tertiary amines, aromatic hydrocarbons). In the third group are compounds whose functional group has both donor and acceptor character such as hydroxyl and amino groups. The remaining two groups include compounds containing at least two of the above-mentioned groups which are distant from one another and form intermolecular bridges, and compounds which do not form a hydrogen bridge at all (e.g. carbon disulphide, carbon tetrachloride, cyclohexane and aliphatic hydrocarbons).

Generally it may be said that the strength of the hydrogen bridge is the greater, the more labile the hydrogen in the molecule of a donor and the higher the electron density in the free pair of the acceptor.

The mutual influence of solvent and compound in chromatography on alumina was studied first on the pair cholesterol and 3β-dimethylaminocholest-5-ene. It had

TABLE 3

THE RELATIONSHIP BETWEEN DONOR-ACCEPTOR PROPERTIES OF SOLVENTS AND MOBILITY OF CHOLESTEROL AND 3β-DIMETHYLAMINOCHOLEST-5-ENE

Solvent system	R_F Values		Adsorptivity
	cholesterol (I)	3β-dimethylamino-cholest-5-ene (II)	
ether	0.77	0.54	II > I
1,2-dichloroethane	0.26	0.10	II > I
chloroform	0.42	0.37	II > I
heptane–10% ethanol	0.20	0.33	I > II
heptane–20% ethanol	0.31	0.56	I > II

TABLE 4
THE MOBILITY OF NITROPHENOLS IN DEPENDENCE ON THE PROPERTIES OF SOLVENTS

Solvent system	R_F Values			Adsorptivity
	o	m	p	
chloroform	0.26	0.03	0.01	$p > m > o$
ether	0.05	0.26	0.03	$p > o > m$
diisopropyl ether	0.18	0.44	0.15	$p > o > m$
benzene–10% ethanol	0.44	0.50	0.20	$p > o > m$

already been found[7] that cholesterol moved faster than the base in a mixture of light petroleum–benzene or in benzene, i.e. in neutral solvents or those forming weak hydrogen bridges. The same sequence was retained also in ether (Table 3). In this solvent which is a hydrogen acceptor, no hydrogen bridge could be formed with the tertiary amine but only with cholesterol which therefore moved faster. A different situation occurred when the chromatogram was developed with a solvent which was a donor. In this case the hydrogen bridge was formed with both chromatographed compounds; it was however stronger with the tertiary amine which is a stronger base than the hydroxyl group. Accordingly, the stronger the solvent as a donor the relatively faster the base moved, finally having a higher R_F value than cholesterol.

A similar relationship has been found with nitrophenols as well. As has been already mentioned, the strength of a hydrogen bridge is proportional to a certain extent to the acidity of the hydrogen donor. It will decrease for nitrophenols as follows: p-nitrophenol (pK_a 7.14), o-nitrophenol (pK_a 7.23) and m-nitrophenol (pK_2 8.35). o-Nitrophenol forms a strong intramolecular bridge and its ability to form an intermolecular bridge is suppressed. It may be expected that the o-derivative will be, in comparison with p- and m-derivatives more weakly bound not only to the adsorbent but to the solvent as well. The position of the o-derivative may therefore serve as a measure of strength of intermolecular bridges formed between m- and p-derivatives and adsorbent on one side and between the solvent on the other. Thus in solvent systems which are donors e.g. chloroform (Table 4), intermolecular bridges between substance and alumina predominate. For this reason, the adsorptivity increases in a sequence o, m, p. In solvent systems which have the character of hydrogen acceptors, the bridges are formed between m- and p-derivatives and the solvent. Thus the o-derivative may be preceded not only by the m-derivative (in ether) but also by the more acid p-derivative in diisopropyl ether which is a stronger hydrogen acceptor than ether.

REFERENCES

1. J. J. PEIFER, Mikrochim. Acta, (1962) 529.
2. S. HUNECK, J. Chromatog., 7 (1961) 561.
3. H. BROCKMANN AND H. SCHODDER, Chem. Ber., 74 (1941) 73.

4. M. Mottier and M. Potterat, *Anal. Chim. Acta*, 13 (1955) 46 and earlier papers.
5. E. A. Mistryukov, *Collection Czech. Chem. Commun.*, 26 (1961) 2071.
6. V. Černy, J. Joska and L. Lábler, *Collection Czech. Chem. Commun.*, 26 (1961) 1658.
7. S. Heřmánek, V. Schwarz and Z. Čekan, *Collection Czech. Chem. Commun.*, 26 (1961) 1669.
8. S. Heřmánek, V. Schwarz and Z. Čekan, *Pharmazie*, 16 (1961) 566.
9. M. Brenner and A. Niederwieser, *Experientia*, 17 (1961) 237.
10. V. Schwarz, Prague, unpublished results.
11. E. A. Mistryukov, *J. Chromatog.*, 9 (1962) 311.
12. E. A. Mistryukov, Moscow, private communication.
13. H. Brockmann and F. Volpers, *Chem. Ber.*, 80 (1947) 77.
14. S. Heřmánek, S. Schwarz and Z. Čekan, *Collection Czech. Chem. Commun.*, 26 (1961) 3170.
15. J. Joska, Prague, unpublished results.
16. D. Waldi, K. Schnackerz and F. Munter, *J. Chromatog.*, 6 (1961) 61.
17. J. Cochin and J. W. Daly, *Experientia*, 18 (1962) 294.
18. A. Mariani and O. Mariani–Marelli, *Rend. Ist. Super. Sanità*, 22 (1959) 759.
19. W. Awe and W. Winkler, *Arzneimittel-Forsch.*, 9 (1959) 773.
20. J. Mokry, L. Dúbravková and P. Šefčovič, *Experientia*, 18 (1962) 564.
21. B. K. Moza and J. Trojánek, *Chem. Ind. (London)*, (1962) 1425.
22. A. Rother, J. M. Bobbit and A. E. Schwarting, *Chem. Ind. (London)*, (1962) 654.
23. R. Tschesche and K. Kometani, *Chem. Ber.*, 94 (1961) 3327.
24. W. Winkler, *Naturwiss.*, 48 (1962) 694.
25. A. Hofmann, *Planta Med.*, 9 (1961) 354.
26. *Aluminium Oxyd für Dünnschichtchromatographie*, Fluka A. G., Buchs.
27. J. Davídek and J. Blattná, *J. Chromatog.*, 7 (1962) 204.
28. D. Waldi, in E. Stahl, *Dünnschicht-Chromatographie*, Springer, Berlin, Göttingen, Heidelberg, 1962, p. 328.
29. Cf. Ref. 28, p. 304.
30. Cf. Ref. 28, p. 477.
31. F. Geiss and H. Schlitt, Euratom, AUR I-1-Chemistry, Nov. 1961; E. Stahl, *Dünnschicht-Chromatographie*, Springer, Berlin, Göttingen, Heidelberg, 1962, p. 378.
32. J. Bäumler and S. Rippstein, *Helv. Chim. Acta*, 44 (1961) 1162.
33. Cf. Ref. 28, p. 321.
34. S. Heřmánek, V. Schwarz and Z. Čekan, *Collection Czech. Chem. Commun.*, in the press.
35. L. Novotný and V. Herout, *Collection Czech. Chem. Commun.*, 27 (1961) 2463.
36. M. Souček, *Collection Czech. Chem. Commun.*, 27 (1962) 2929.
37. M. Suchý, V. Herout and F. Šorm, *Collection Czech. Chem. Commun.*, 27 (1962) 2398.
38. J. Vrkoč, V. Herout and F. Šorm, *Collection Czech. Chem. Commun.*, in the press.
39. G. Schetty and W. Kuster, *Helv. Chim. Acta*, 44 (1961) 2193.
40. J. Davídek and G. Janíček, *Z. Lebensm.-Untersuch. Forsch.*, 115 (1961) 113.
41. J. Rosmus and Z. Deyl, *J. Chromatog.*, 6 (1961) 187.
42. W. Trappe, *Biochem. Z.*, 305 (1940) 150.
43. H. H. Strain, *Chromatographic Adsorption Analysis*, Interscience, New York, 1942.
44. J. Jacques and J. P. Mathieu, *Bull. Soc. Chim. France*, (1946) 94.
45. J. v. Euw and T. Reichstein, *Helv. Chim. Acta*, 31 (1948) 883.
46. T. Reichstein and C. W. Shoppee, *Discussions Faraday Soc.*, 7 (1949) 305.
47. R. Neher, *Chromatographie von Sterinen, Steroiden und verwandten Verbindungen*, Elsevier, 1958.
48. E. Hecker, *Chimia (Aarau)*, 8 (1954) 233.

LA CHROMATOGRAPHIE SUR COUCHES MINCES DANS LE DOMAINE DES SUBSTANCES ODORANTES NATURELLES ET SYNTHÉTIQUES

E. DEMOLE

Firmenich et Cie, Genève (Suisse)

1. INTRODUCTION

Évoquer ici même l'utilité de la chromatographie sur couches minces dans le domaine des substances odorantes nous paraît être une tâche agréable et toute naturelle. Agréable, car nous nous trouvons devant un auditoire averti et pleinement familiarisé avec cette nouvelle technique, ce qui évite de devoir procéder aux digressions accessoires qui seraient nécessaires dans des circonstances moins favorables. Je remercie à ce propos le Professeur STAHL et tous les orateurs qui ont brillamment ravivé notre connaissance des principes généraux de la chromatographie sur couches minces. Le présent exposé constitue aussi une tâche toute naturelle puisque ce type de chromatographie a connu ses premières applications précisément dans le domaine des substances odorantes, quand KIRCHNER et ses collaborateurs[1] réalisèrent leurs chromatostrips en 1951, puis REITSEMA[2] ses chromatoplates 3 années plus tard. Il serait toutefois trop monotone de suivre simplement la voie ouverte par ces auteurs et de nous borner à la description chronologique des diverses améliorations qui ont succédé à leurs travaux. D'ailleurs, ce genre de compilation existe déjà dans la littérature, tout particulièrement depuis la parution des ouvrages spécialisés de STAHL, de RANDERATH et de TRUTER. Nous aimerions plutôt mettre en évidence certaines qualités particulières de la chromatographie sur couches minces et montrer de quelle manière ces qualités peuvent faciliter l'analyse chimique totale des huiles essentielles. Nous parlerons ensuite d'une adaptation particulière des chromatoplaques à l'étude des corps odorants, puis nous tenterons d'effleurer le très intéressant problème des relations entre le comportement chromatographique et la structure moléculaire des substances.

L'étude des constituants des huiles essentielles offre un double intérêt académique et industriel. D'une part, elle conduit à une meilleure connaissance du mode d'action de la nature dans l'élaboration des structures organiques, et il suffira à ce propos de rappeler que la grande chimie des terpènes a trouvé son origine dans de telles études. D'autre part, sur le plan industriel, l'identification des éléments odorants caractéristiques d'une essence naturelle permet souvent d'améliorer ou d'imiter celle-ci, ou autorise la création de mélanges odorants nouveaux. Enfin, dans le cas idéal, toujours

Bibliographie p. 54

très rare, d'une analyse suffisamment poussée, il devient en principe possible de procéder à la reconstitution synthétique exacte et totale de l'essence naturelle. Mais ce genre de reconstitution intégrale d'huiles essentielles, extraordinairement délicat à réaliser, nécessite à tout le moins le concours de l'ensemble des techniques analytiques modernes, réunies en une judicieuse association. Dans le meilleur des cas, le chimiste ne peut éviter de devoir confier sa préparation finale à un parfumeur qui cherchera, au moyen d'essais empiriques patients, à pallier les effets des petites insuffisances de l'analyse. En effet, certaines nuances secondaires mais importantes dans l'odeur des essences naturelles dépendent souvent de constituants présents en traces et qui échappent au moins partiellement à l'analyse la plus rigoureuse. Celle-ci, dans ce domaine, n'est donc jamais que relativement complète, et l'intervention finale du parfumeur reste aujourd'hui encore nécessaire pour parachever l'œuvre du chimiste.

Nous sommes entré en contact avec ce genre de problème il y a 8 ans, en abordant l'analyse totale de l'essence de jasmin dans le but de la reconstituer ensuite synthétiquement. Cette essence possède un certain intérêt industriel, elle a déjà été analysée à plusieurs reprises depuis le début du siècle, mais l'ensemble des résultats publiés en 1955 ne permettait aucunement de la reconstituer. Il restait visiblement à découvrir certains constituants importants, et nous avons décidé de les rechercher au moyen d'une technique d'analyse basée entièrement sur l'emploi de la chromatographie, contrairement aux méthodes chimiques plus drastiques adoptées par nos prédécesseurs. Dans ces conditions, les publications originales de KIRCHNER et ses collaborateurs[1] et de REITSEMA[2], récentes à l'époque, devaient obligatoirement attirer notre attention, et nous avons tenu à introduire l'emploi des chromatostrips et chromatoplates dans notre laboratoire. Ces procédés ne nous semblaient d'ailleurs pas offrir en eux-mêmes une très grande utilité pour l'étude directe des huiles essentielles, mais ils pouvaient par contre rendre de grands services comme moyens de contrôle dans la mise au point, l'observation et l'interprétation des fractionnements préparatifs sur colonnes, possibilité déjà suggérée par MILLER ET KIRCHNER[1c]. En 1955, la chromatographie sur couches minces était pratiquement inconnue en Europe et, bien entendu, la technique développée par le Professeur STAHL[3] n'existait pas, de sorte qu'il nous a fallu près d'un mois d'efforts avant d'être capable de préparer avec régularité des chromatoplaques utilisables. Mais nous avons été amplement récompensé de cette persévérance, et nous allons maintenant décrire notre méthode d'analyse de l'essence de jasmin, qui pourrait bien constituer la base d'une systématique généralement applicable à l'étude des huiles essentielles et des extraits végétaux.

2. TECHNIQUE ET RÉSULTATS

Nous avons préalablement préparé l'essence de jasmin pour l'examen chromatographique en la débarrassant, par distillation, de ses constituants les plus volatils ainsi que des parties indistillables non intéressantes. La fraction d'essence bouillant de 100 à 160° sous 0,05 mm Hg retenue pour nos travaux a été ensuite examinée méthodiquement sur chromatostrips siliciques.

Dans une première étape, nous avons recherché et trouvé le système de solvants convenant le mieux à la séparation des divers constituants présents.

Dans une seconde étape, nous avons précisé la nature de ces constituants en révélant une série de chromatostrips avec des réactifs spécifiques des fonctions cétone, aldéhyde, phénol, etc.

Dans une troisième étape, les zones correspondant aux constituants révélables chimiquement ont été prélevées sur un chromatostrip non révélé, puis éluées au moyen d'un solvant polaire, et ensuite étudiées et classées du point de vue de l'odeur.

Dans une quatrième étape, nous avons réalisé la séparation micro-préparative d'environ 1 mg d'essence sur un chromatostrip spécialement épais puis, après développement, prélevé et élué les zones intéressantes. Un examen direct des éluats par spectrophotométrie ultra-violette a permis de mettre en évidence plusieurs chromophores caractéristiques tels que cétone $\alpha\beta$-insaturée, phénol, diène conjugué, etc.

Cette étude préliminaire sur chromatostrips, qui n'a coûté que quelques mg d'essence de jasmin, a permis d'établir la présence de 8 groupes de constituants séparables chromatographiquement sur acide silicique, et de déceler, dans ces 8 groupes, des substances odorantes de nature lactonique, cétonique, phénolique et indolique, accompagnées d'hydrocarbures, d'esters et d'alcools peu odorants. Elle a encore montré que l'odeur typique du jasmin était due aux constituants cétoniques et, bien entendu, elle nous a permis aussi de définir les conditions à adopter pour réaliser un fractionnement préparatif correct de l'essence de jasmin. Voilà qui illustre la très grande utilité de la chromatographie sur couches minces pour l'étude préliminaire et exploratoire d'une huile essentielle.

Nous pouvions après cela aborder avec confiance les séparations préparatives sur colonnes, effectuées en présence de benzène seul ou mélangé d'acétate d'éthyle, et avec une quantité d'adsorbant égale à $20 \times$ celle de l'essence. Chaque fraction obtenue a été immédiatement comparée, sur chromatoplaques, avec l'essence initiale non chromatographiée. Ces contrôles nous ont permis de suivre très facilement la progression générale du fractionnement chromatographique préparatif, et de vérifier constamment la pureté des fractions obtenues. Inévitablement, certaines de celles-ci ont dû subir une purification chromatographique complémentaire car les séparations sur colonnes sont généralement moins nettes que sur couches minces. Finalement, nous avions en notre possession une série de fractions chromatographiques dont le poids variait de plusieurs centaines de mg à plusieurs g, et qui correspondaient fidèlement aux divers constituants décelés lors de l'étude préliminaire de l'essence de jasmin sur chromatostrips. Un résultat aussi précis n'aurait pu être atteint sans l'utilisation constante de la chromatographie sur couches minces pour le contrôle du fractionnement préparatif. L'étude de ces produits a montré que la plupart d'entre eux étaient encore des mélanges, mais leur résolution finale n'a pas posé de problème

Bibliographie p. 54

important et a pu être réalisée en recourant soit à la distillation, à la chromatographie en phase gazeuse, ou même à une seconde chromatographie silicique effectuée dans des conditions spécialement étudiées. Leur analyse chimique[4] a démontré que la fraction d'essence de jasmin bouillant de 100 à 160° sous 0,05 mm Hg se compose, dans l'ensemble, d'hydrocarbures insaturés, de benzoate de benzyle, de benzoate de $\beta\gamma$-hexényle, de linolénate de méthyle, de palmitate de méthyle, d'acétate de phytyle, de phytol, d'isophytol, de géranyl-linalol, d'eugénol, de jasmone, de jasmonate de méthyle, de lactone de l'acide hydroxy-5-décène-7-oïque, de N-acétylanthranilate de méthyle et d'indol. Tous ces constituants ont été distingués ou isolés uniquement par voie physico-chimique, en grande partie grâce à la participation de la chromatographie sur couches minces. Un certain nombre d'entre eux étaient nouveaux à l'époque, et il a fallu étudier leur structure et réaliser leur synthèse.

3. Premières conclusions

En premier lieu, soulignons que la chromatographie silicique permet de réaliser, dans des conditions infiniment plus douces que par la méthode chimique classique, une véritable analyse fonctionnelle de l'essence de jasmin, et sans doute ceci est-il valable pour d'autres huiles essentielles. Les constituants cétoniques, phénoliques, les esters, les alcools se séparent nettement par cette méthode et n'interfèrent entre eux que d'une manière occasionnelle et peu gênante. La fraction cétonique isolée de l'essence de jasmin par la voie chromatographique que nous avons décrite se compose par exemple seulement de jasmone et de jasmonate de méthyle aisément séparables par distillation fractionnée. On peut isoler cette même fraction cétonique par voie chimique, en traitant l'essence de jasmin par le réactif «P» de Girard & Sandulesco[5], mais dans ce cas le rendement est plus faible et surtout il ne serait plus possible de continuer en toute sécurité l'analyse de l'essence ainsi traitée, par suite des altérations qu'elle aurait pu subir.

En second lieu, remarquons que tous les avantages procurés par l'emploi de la chromatographie silicique sur colonnes, dans le domaine des huiles essentielles, ne peuvent être pleinement exploités que si on lui associe continuellement la chromatographie sur couches minces, indispensable pour l'étude préliminaire, la mise au point des séparations et leur contrôle. Cette condition n'est d'ailleurs aucunement limitative et, dans certains cas, on peut être amené à adjoindre encore à cette combinaison la chromatographie en phase vapeur ou toute autre technique adéquate.

En troisième lieu enfin, il est intéressant de relever que, dans le cas de l'essence de jasmin, la chromatographie sur couches minces a permis de résoudre rapidement le problème posé et de ce fait a directement contribué à la réalisation d'un objectif industriel.

4. FRACTIONNEMENT PAR DIFFUSION ENTRE CHROMATOPLAQUES

Nous désirons aborder maintenant notre deuxième sujet concernant une adaptation particulière de la chromatographie sur couches minces à l'étude des matières odorantes. On sait qu'en règle générale les substances odorantes possèdent une polarité modérée alliée à une certaine volatilité sans laquelle leur diffusion dans l'atmosphère, et par conséquent leur contact avec les terminaisons nerveuses de notre muqueuse olfactive, serait impossible. Cette volatilité indispensable varie d'ailleurs dans des limites très larges suivant le point d'ébullition et la structure de ces substances, suivant le fait qu'elles se trouvent à l'état pur ou au contraire mélangées à d'autres corps odorants ou non, et suivant leur concentration moléculaire. Tout mélange odorant d'une certaine complexité, nous dirons tout parfum, est finalement caractérisé par une courbe d'évaporation composite qui reflète l'influence globale des diverses associations intermoléculaires des constituants sur la volatilité individuelle de chacun d'eux. Ce comportement à l'évaporation est un facteur très important pour la qualité des parfums, car l'odeur de ceux-ci ne doit pas perdre trop rapidement de sa cohésion, ni se transformer trop aisément en s'éventant. Les parfumeurs professionnels ont pour coutume d'étudier ce facteur d'une manière empirique mais efficace, en abandonnant simplement à l'air une touche de papier spécial, imprégnée de parfum à l'une de ses extrémités, puis en surveillant l'évolution de l'odeur avec le temps.

Nous nous sommes demandé si, dans ce domaine qui est en somme celui de l'analyse immédiate des parfums, il ne serait pas possible de procéder plus scientifiquement grâce à l'emploi des couches minces adsorbantes. Notre idée de base était d'essayer d'utiliser la volatilité propre aux substances odorantes comme facteur de séparation supplémentaire, c'est-à-dire de combiner la diffusion gazeuse et l'adsorption. Après un certain nombre d'essais préliminaires, nous avons retenu la méthode suivante.

On imprègne aussi régulièrement que possible l'ensemble de la surface d'une chromatoplaque silicique avec une solution éthéro-pétrolique du mélange odorant. Une plaque de 20 $\times$ 5 cm peut par exemple recevoir une quantité de l'ordre de 100 mg de substance. Après évaporation du solvant, on recouvre la plaque imprégnée d'une seconde chromatoplaque neuve, de telle sorte qu'une distance de l'ordre du mm sépare les deux couches adsorbantes placées en regard l'une de l'autre. Dans ces conditions, les corps odorants diffusent peu à peu de la première couche vers la seconde où ils se fixent, et la rapidité du phénomène dépend de leur point d'ébullition, de leur affinité pour l'adsorbant, et, bien entendu, des conditions pratiques adoptées. L'opération peut être réalisée sous un vide plus ou moins poussé ou à une certaine température afin d'accélérer la diffusion des constituants moyennement ou peu volatils. Nous avons fréquemment observé qu'une heure peut suffire à assurer le transfert gazeux de 1 à 20 mg, quelquefois 40 mg, de substance d'une couche adsorbante à l'autre, ceci dans le cas des plaques de 20 $\times$ 5 cm. On peut bien entendu réaliser un fractionnement méthodique des mélanges odorants en remplaçant à plusieurs reprises la plaque réceptrice tout en poussant progressivement les conditions de la diffusion. L'isolement final des diverses fractions s'effectue très facilement en éluant l'adsorbant

Bibliographie p. 54

des plaques réceptrices, disposé sous forme de petites colonnes, avec un solvant convenable. L'analyse peut ensuite être continuée avec profit en examinant ces fractions par chromatographie sur couches minces habituelle, par chromatographie en phase vapeur ou par une autre technique appropriée.

Ce procédé d'étude des substances odorantes est presque aussi «naturel» que la classique méthode à la touche des parfumeurs. Mais, en plus d'une précision accrue, il présente sur celle-ci le grand avantage de permettre aussi bien l'examen des constituants qui se sont évaporés que des fractions résiduelles. Son efficacité, qui fait l'objet d'études dans notre laboratoire, peut être appréciée dans une certaine mesure si l'on considère les résultats des tests préliminaires réalisés sur des mélanges binaires de substances odorantes connues. C'est ainsi qu'un mélange de géraniol et de linalol en parties égales se sépare, au cours de la diffusion entre deux chromatoplaques, en donnant une fraction volatile titrant plus de 80% de linalol, une fraction résiduelle titrant plus de 90% de géraniol, et des mélanges intermédiaires. Le fractionnement est relativement efficace car le constituant le moins volatil du mélange, le géraniol, est aussi dans ce cas le plus polaire, et l'effet de diffusion s'ajoute donc à l'effet d'adsorption pour promouvoir la séparation. Il n'en est pas de même en ce qui concerne la paire jasmone + jasmonate de méthyle car ces substances présentent sensiblement la même affinité pour les couches siliciques, et leur séparation ne peut guère s'effectuer que grâce à la différence de leurs points d'ébullition. Nous avons néanmoins obtenu, dans ce cas, une fraction volatile contenant 94% de jasmone et une fraction résiduelle titrant 82% de jasmonate de méthyle, en partant toujours d'un mélange en proportions égales. Notre troisième exemple de séparation sera celui de la paire acétate de benzyle + linalol, où ce dernier est à la fois le plus volatil et le plus polaire des deux constituants, ce qui oppose l'effet de diffusion à l'effet d'adsorption au cours du fractionnement. C'est l'effet d'adsorption qui reste déterminant car nous avons obtenu une première fraction titrant 73% d'acétate de benzyle et une fraction résiduelle titrant 90% de linalol, en dépit du fait que ce dernier soit le plus volatil des deux constituants. Nous avons encore étudié la séparation des α- et β-ionones, également à partir d'un mélange en proportions égales. Le fractionnement s'est révélé moins complet dans ce cas que dans les précédents, ce qui est assez normal car ces isomères ont un point d'ébullition et un comportement chromatographique très semblables. Nous avons obtenu une première fraction titrant 62% d'α-ionone et une fraction résiduelle titrant 60% en isomère β. Il serait bien entendu possible de réaliser des fractionnements plus précis et d'obtenir des substances probablement pures en reprenant les fractions extrêmes et en les soumettant à une ou plusieurs nouvelles séparations par diffusion.

Nous avons cité ces quatre tests préliminaires dans le seul but de fixer les idées et de démontrer la réalité pratique du fractionnement des matières volatiles par diffusion entre deux chromatoplaques. Il est évident que l'utilité de la méthode ne réside aucunement dans la séparation partielle des mélanges binaires précédents, qui pourraient être étudiés bien plus efficacement par chromatographie en phase vapeur, mais dans l'étude de mélanges beaucoup plus complexes comme les huiles essentielles par exemple. L'analyse de celles-ci nécessite en effet souvent l'exécution d'un fractionnement chimique ou physique préalable pour concentrer dans certaines fractions les éléments odorants à la fois importants et présents en petite quantité. La diffusion entre chromatoplaques représente un moyen d'effectuer cette opération à partir de peu de substance et dans des conditions particulièrement douces. Nous avons par exemple effectué un cinquième test portant sur un échantillon de 100 mg d'essence absolue de jasmin et recueilli 6 fractions volatiles successives. Leur étude par chromatographie en phase gazeuse a permis de détecter facilement les constituants les plus caractéristiques du jasmin, notamment l'acétate de benzyle, le linalol, la jasmone, etc., concentrés électivement dans certaines d'entre elles.

En résumé, l'emploi classique des chromatoplaques dans l'étude des mélanges odorants pourrait quelquefois être complété avec profit par la technique de diffusion que nous venons de décrire. Cette méthode, qui représente en soi une variante particulière de séparation à l'échelle micro-préparative, autorise le fractionnement des mélanges en l'absence de tout chauffage, contrairement à la distillation et à la chromatographie en phase vapeur. Elle présente l'intérêt évident de fractionner les corps odorants par classes de volatilité et de polarité, de fournir au chimiste un moyen de procéder à la séparation préliminaire de petites quantités de mélanges chimiquement délicats, et au parfumeur un procédé d'étude plus précis des parfums. Son domaine exact d'application sera précisé ultérieurement par les études en cours dans notre laboratoire.

5. STRUCTURE MOLÉCULAIRE ET COMPORTEMENT CHROMATOGRAPHIQUE

L'introduction et le développement de la chromatographie sur couches minces devraient faciliter et stimuler l'étude des relations existant entre le comportement chromatographique et la structure chimique des substances, pour les deux raisons suivantes. Tout d'abord, l'emploi de cette technique simplifie dans une très grande mesure la détermination et la comparaison précises des valeurs de R_F dans un processus d'adsorption, et représente un important progrès par rapport aux colonnes que LeRosen et ses collaborateurs[6] ont utilisées dans le même but en 1948. Ensuite, l'interprétation du comportement chromatographique général des substances se prête à une simplification séduisante dans le cas du silicagel ou de l'acide silicique communément employés sur chromatoplaques. On sait en effet que ce type d'adsorbant agit principalement grâce à la formation de liaisons hydrogène entre ses groupes silanols et les fonctions polaires du soluté, ce qui ramène en définitive le processus chromatographique à un simple échange électronique entre les deux parties, du moins si l'on néglige *a priori* l'intervention possible d'autres mécanismes secondaires. Ce principe permet de penser que l'affinité de base d'une molécule donnée pour le silicagel dépend avant tout de la polarité du ou des groupes fonctionnels présents et de leur degré d'empêchement stérique, et qu'il serait donc possible de prévoir le comportement chromatographique relatif d'une substance quelconque si l'on parvenait à chiffrer d'une manière assez précise ces deux facteurs. Malheureusement, le problème se révèle d'autant plus compliqué que, suivant les cas, les effets électroniques et stériques peuvent s'exercer indépendamment les uns des autres ou se conjuguer. Par exemple, le fait qu'un alcool tertiaire soit normalement moins adsorbé que son isomère primaire s'explique aussi bien par l'effet donneur d'électrons des substituants hydrocarbonés du premier, ce qui diminue l'acidité de l'hydroxyle, que par l'empêchement stérique apporté par ces mêmes substituants. Les deux effets sont ici superposés. Par contre, la légère différence observable entre les propriétés chromatographiques des β- et α-ionones s'explique principalement par l'intervention d'un effet électronique dû à l'allongement du chromophore conjugué du premier isomère, qui devient de ce fait plus polaire. Dans cet exemple, c'est donc l'effet électronique qui prédomine. Enfin, si nous considérons le cas d'alcools stéréoisomères comme les menthols, qui ne diffèrent que par leur conformation ou l'orientation spatiale de l'hydroxyle, nous nous trouvons dans la situation où seuls des facteurs stériques peuvent expliquer les différences observées dans le comportement chromatographique.

Il faut reconnaître qu'aujourd'hui encore il n'est pas possible d'évaluer d'une manière précise et générale l'influence respective des effets stériques et électroniques sur le comportement chromatographique des molécules. Le plus souvent, on doit se borner à estimer expérimentalement ces facteurs et à les inclure sous forme de constantes empiriques, par exemple dans le calcul des valeurs de R_M[7].

Nous nous sommes toutefois demandé s'il ne serait pas possible de parvenir à établir une relation plus ou moins directe entre les propriétés chromatographiques et certains paramètres intramoléculaires, dans des cas particulièrement favorables. Les alcools cyclaniques, de conformation rigide ou suffisamment «figée» nous ont paru devoir constituer un matériel idéal pour une telle étude, car leur comportement chromatographique dépend presque uniquement d'effets stériques relativement faciles à évaluer si l'on dispose d'un modèle moléculaire précis et d'une méthode de calcul appropriée. Nous avons utilisé le modèle moléculaire de DREIDING, sur lequel 1 cm correspond à 0,4 Å, et admis arbitrairement, pour nos calculs, que l'empêchement stérique du groupe OH devait être égal à la somme des inverses des carrés des distances qui le séparent de chacun des atomes de carbone de la molécule; ce postulat est purement artificiel et ne prétend nullement représenter un phénomène réel, mais il permet de chiffrer approximativement et d'une manière comparable le degré d'empêchement stérique de la fonction alcool dans un groupe donné de stéréoisomères. Illustrons par exemple son application à l'étude des menthols.

Dans une première étape nous reproduisons, à l'aide du modèle moléculaire, les quatre stéréoisomères correspondants, c'est-à-dire les menthol, isomenthol, néomenthol et néoisomenthol; on sait que ces alcools diffèrent seulement par l'orientation des groupes méthyle et hydroxyle, le groupe isopropyle restant par définition équatorial. Dans une seconde étape, nous mesurons, directement sur les quatre modèles moléculaires, les distances qui séparent en ligne directe l'atome d'oxygène de chacun des atomes de carbone constituant le squelette rigide de la molécule; les deux méthyles du groupe isopropyle, qui peuvent osciller, sont donc négligés dans ces mesures. Enfin, nous incluons les distances ainsi mesurées dans une équation simple basée sur le postulat déjà exprimé, et nous obtenons pour chaque menthol un coefficient caractéristique d'empêchement stérique E.

$$E = \sqrt{\dfrac{1}{1/D_a{}^2 + 1/D_b{}^2 + 1/D_c{}^2 + \ldots}}$$

$D_a, D_b, D_c, \ldots$ = distances interatomiques mesurées en cm, sur les modèles, entre l'oxygène du groupe —OH et chacun des atomes de carbone constituant le squelette rigide de la molécule.

D'après le mécanisme admis de l'adsorption sur l'acide silicique, il devrait exister un rapport direct entre ces coefficients et les propriétés chromatographiques des menthols. Tel est bien le cas car le calcul montre que l'empêchement stérique du groupe OH croît dans l'ordre menthol $<$ isomenthol $\ll$ néomenthol $<$ néoisomenthol, ce qui correspond assez exactement à la distribution des valeurs de R_F observées sur chromatoplaques par PETROWITZ[8] en présence de benzène:

	menthol	isomenthol	néomenthol	néoisomenthol
$E =$	2,26	2,25	2,15	2,08
$R_F =$	0,16	0,17	0,28	0,29
	0,36	0,37	0,51	0,55*

* R_F obtenus en présence de 5 % de méthanol.

Le même calcul, appliqué en série stéroïde, permet de prévoir que le cholestanol doit être beaucoup plus polaire que l'épicholestanol et le coprostanol, et que ces deux derniers ne doivent guère différer entre eux. Cette prévision, confirmée par les résultats expérimentaux, s'accorde parfaitement avec la généralisation de NEHER[9] qui admet que le comportement chromatographique des stéroïdes hydroxylés en 3- dépend avant tout de la conformation de l'hydroxyle plutôt que du mode de soudure *cis* ou *trans* des cycles A et B.

cholestanol	épicholestanol	coprostanol
$E = 2,15$	2,02	1,99

Une prévision approchée des propriétés chromatographiques relatives d'alcools stéréoisomères semble donc permise dans une certaine mesure par l'emploi des mo-

Bibliographie p. 54

dèles moléculaires. Mais cette possibilité reste encore empirique dans son état actuel, et ne fait que préciser certains concepts déjà bien connus, tel celui qui attribue une polarité généralement plus grande aux alcools équatoriaux qu'aux épimères axiaux. Nous pensons que cette méthode pourrait toutefois se révéler utile dans la prévision du comportement chromatographique de substances différant par plus d'une cause de stéréoisomérie, et dans leur étude conformationnelle d'après leur ordre d'élution sur chromatoplaques siliciques. Dans cette dernière perspective, elle pourrait compléter et recouper avec profit la technique connue basée sur la comparaison des valeurs ΔR_M entre séries isomères.

Bien entendu, le calcul des coefficients stériques ne peut avoir de sens que sur des substances conformationnellement stables et de même structure électronique, engagées dans un mécanisme d'adsorption par liaisons hydrogène, c'est-à-dire avant tout dans le cas des couches minces à base de silicagel et en chromatographie de partage sur papier. D'autre part, on doit s'attendre à observer quelquefois certaines irrégularités, par exemple lorsque les valeurs de R_F sont déterminées en présence de solvants très polaires susceptibles de s'associer avec les substances chromatographiées. Enfin, la méthode d'estimation des coefficients stériques doit être encore complétée de manière à tenir également compte de l'influence des atomes d'hydrogène pour fournir ainsi des résultats plus précis.

Malgré ces dernières réserves, nous espérons avoir montré que l'emploi des modèles moléculaires n'est pas sans intérêt dans la prévision du comportement des substances en chromatographie sur couches minces. Nous terminons en formant nos meilleurs vœux pour que les nombreux utilisateurs actuels de cette élégante technique accordent, à l'avenir, une attention accrue au problème fondamental des relations entre structure moléculaire et comportement chromatographique.

BIBLIOGRAPHIE

1. a J. G. KIRCHNER, J. M. MILLER ET G. J. KELLER, Anal. Chem., 23 (1951) 420.
   b J. M. MILLER ET J. G. KIRCHNER, Anal. Chem., 25 (1953) 1107.
   c J. M. MILLER ET J. G. KIRCHNER, Anal. Chem., 24 (1952) 1480.
   d J. G. KIRCHNER ET J. M. MILLER, Ind. Eng. Chem., 44 (1952) 318.
   e J. G. KIRCHNER, J. M. MILLER ET R. G. RICE, J. Agr. Food Chem., 2 (1954) 1031.
   f J. M. MILLER ET J. G. KIRCHNER, Anal. Chem., 23 (1951) 428.
2. R. H. REITSEMA, Anal. Chem., 26 (1954) 960; J. Am. Pharm. Assoc., Sci. Ed., 43 (1954) 414.
3. E. STAHL, Chemiker Ztg., 82 (1958) 323.
4. E. DEMOLE, Compt. Rend., 243 (1956) 1883; E. DEMOLE ET E. LEDERER, Bull. Soc. Chim. France, (1958) 1128; E. DEMOLE, Thèse, série A, no. 844, no. d'ordre 870, Paris, 1958; E. DEMOLE, E. LEDERER ET D. MERCIER, Helv. Chim. Acta, 45 (1962) 675; 45 (1962) 685; E. DEMOLE ET M. STOLL, Helv. Chim. Acta, 45 (1962) 692; M. WINTER, G. MALET, M. PFEIFFER ET E. DEMOLE, Helv. Chim. Acta, 45 (1962) 1250; E. DEMOLE ET M. WINTER, Helv. Chim. Acta, 45 (1962) 1256.
5. E. DEMOLE, Helv. Chim. Acta, 45 (1962) 1951; Y. R. NAVES, A. V. GRAMPOLOFF ET E. DEMOLE, Helv. Chim. Acta, 46 (1963) 1006.
6. A. L. LEROSEN ET A. MAY, Anal. Chem., 20 (1948) 1090; E. D. SMITH ET A. L. LEROSEN, Anal. Chem., 23 (1951) 732; 26 (1954) 928.
7. Voir par exemple: E. STAHL, Dünnschicht-Chromatographie, Springer Verlag, Berlin, Göttingen, Heidelberg, 1962, p. 105.
8. H. J. PETROWITZ, Angew. Chem., 72 (1960) 921.
9. R. NEHER, Chromatographic Reviews, Vol. I, Elsevier, Amsterdam, 1959, p. 185.

THIN-LAYER ELECTROPHORESIS

GIULIANA GRASSINI

Istituto di Chimica Generale ed Inorganica dell' Università, Laboratorio di Chimica delle Radiazioni e Chimica Nucleare del C.N.E.N., Rome (Italy)

1. INTRODUCTION

Although thin-layer electrophoresis has been known for many years, it has not as yet found such wide application as the analogous technique of thin-layer chromatography. As early as 1946, CONSDEN, GORDON AND MARTIN[1] carried out a satisfactory separation of a mixture of amino acids and peptides by electrophoresis in a thin layer of silica gel. The technique used presented no particular practical difficulties, but it certainly could not compete with the extreme simplicity of paper electrophoresis, which was developed at roughly the same time. This may explain why applications for thin-layer electrophoresis[3-7] were only discovered after a space of more than ten years, when the preparation of thin layers had been greatly simplified by STAHL's technique[2].

Just as thin-layer chromatography was in some respects superior to chromatography on paper, thin-layer electrophoresis presented certain advantages over paper electrophoresis: in certain cases separations which were impossible with paper electrophoretic techniques could be carried out, the times of separation were often shorter, and the spots were sharper. Thin-layer electrophoresis was shown to be capable of wide application, and has in fact been used to separate amines, amino acids, phenols, and dyestuffs[3-5], and for the separation of inorganic ions[6,7]. The technique is practically the same as that for paper electrophoresis, except that a thin layer of an adsorbent, supported on a glass plate and prepared by methods similar to those employed in thin-layer chromatography, is used in place of a strip of filter paper.

2. PREPARATION AND COMPARISON OF THIN LAYERS OF THE ADSORBENT

Various types of adsorbent are used in the preparation of thin layers for use in electrophoresis. WIELAND AND PFLEIDERER[8] have described strips made of starch, but the results depended too much on the quality of the adsorbing compound. MARTEN[9] and PFRUNDER et al.[6] used thin layers of agar, the former mainly for biological work and the latter for the separation of inorganic ions. PASTUSKA AND TRINKS[3,4] reported separations on silica gel G and kieselguhr G, whilst HONEGGER[5] tried the same adsorbents and alumina G. Plaster of Paris strips were used by DOBICI AND GRASSINI[7].

The various authors prepared their thin layers in much the same manner. Two

References p. 68

strips of paper (or of shredded paper) were glued to the ends of a glass plate to act as bridges between the thin layer of adsorbent and the electrolyte. The glass plate was then coated to a thickness of 0.4–2 mm with a paste of the adsorbing material and water or the buffer solution used as the electrolyte. Slightly different techniques are used at this stage. The resulting layer was then dried, in air or in an oven, for a definite time which is e.g. 25 min at ambient temperature for silica gel G, and 48 h at 60°C for plaster of Paris.

The characteristics of adsorbents used for the preparation of thin layers are naturally different. HONEGGER attempted to carry out a relative assessment of the more commonly used adsorbents, i.e. silica gel G, kieselguhr G, alumina G, and Whatman paper No. 1. Such comparison was however very difficult, since various factors had to be taken into account. The pH of the electrolyte, for instance, which was initially 3.8, changed to a different extent on the various strips (silica gel pH $\simeq$ 3; kieselguhr pH $\simeq$ 2.5; alumina pH $\simeq$ 4.5; paper pH $\simeq$ 3.5). The moisture contents of the various strips are different and cannot even be determined accurately, leading to differences in diffusion and in the passage of electric current. Furthermore, the sensitivities of colour reactions used to identify the separated substances are also different on different strips.

By comparing the rates of migration of his samples (cysteine, mescaline, cadaverine, and methylamine) HONEGGER was nevertheless able to show that the migrations on kieselguhr and paper were faster than on silica gel and alumina. It was also found that the sizes of spots increased in the series alumina < kieselguhr < silica gel < paper. Taking all characteristics into account, HONEGGER concluded that kieselguhr G was the best material.

3. APPARATUS AND TECHNIQUES EMPLOYED IN ELECTROPHORESIS

The apparatus used for thin-layer electrophoresis may be the same as that used in paper electrophoresis, with or without a cooling system. An apparatus specially constructed for thin-layer electrophoresis (Fig. 1) has recently been placed on the market by Margraf of Berlin. This apparatus is fitted with a rectifier which can supply a maximum voltage of 480 V and a current of up to 50 mA. The $335 \times 140 \times 50$ mm electrophoresis chamber is made of Trovidur, so that organic substances which attack this material cannot be used. Up to five chambers may be connected to the rectifier at the same time. The electrodes are made of platinum and are fixed inside reservoirs which contain the electrolyte. The middle portion, which separates the electrolyte reservoirs, supports the glass plates and may be cooled by passing cold water through the lower part. The entire chamber is closed by a thick glass plate, which may be replaced by a water-cooled slab of glass.

Before electrophoresis is commenced, the strip, which is prepared as described above, is uniformly moistened with the electrolyte to be used in the separation. As in paper electrophoresis, the samples are then applied onto the thin layer by means of a micropipette, in a predetermined position, in the form of small spots or a line

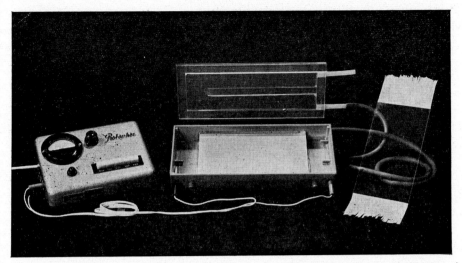

Fig. 1. Apparatus for thin-layer electrophoresis. On the left: rectifier. Centre: electrophoresis chamber. On the right: glass support ready for coating with adsorbent.

which must be as uniform as possible. Fig. 2 shows the results obtained by electrophoresis using increasing quantities (1–20 μg) of syringic acid. It is clear that the electrophoretic displacement is independent of the quantity of syringic acid used.

The time required for the experiment depends on the problems of separation, but in general does not exceed 2 h. At the end of the separation the end strips of paper are detached, and the sample components are developed with appropriate reagents.

To allow a true comparison of migration distances obtained with different substances after various periods of electrophoresis, PASTUSKA and HONEGGER referred

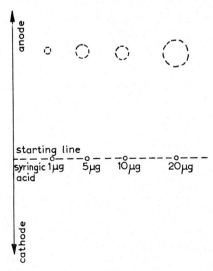

Fig. 2. Electrophoresis with increasing quantities of syringic acid (1–20 μg).

References p. 68

their results to the displacement of an arbitrary standard compound. Such relative values were respectively denoted as M_G and E_f by the two authors. Taking the M_G value of the reference compound as unity, the relative M_G value of the test substance is given by:

$$M_G = \frac{\text{distance travelled by the test substance}}{\text{distance travelled by the reference compound}}$$

M_G is greater than 1 when the test substance migrates at a greater speed than the reference compound.

4. COMPOUNDS EXAMINED

Phenols

This class of compounds is one of the most widely studied by thin-layer electrophoresis. In their first investigations, PASTUSKA AND TRINKS[3] studied the electrophoretic behaviour of 36 phenols on strips of silica gel G and kieselguhr G. The electrolyte used was a mixture of EtOH (80 ml), H_2O (30 ml), H_3BO_3 (4 g), and sodium acetate (2 g). Acetic acid was added to the electrolyte to bring the pH to 4.5 in the case of silica gel, and to 5.5 in the case of kieselguhr. An important difference between paper and thin-layer electrophoresis is that in the former the separation depends strongly on pH, whilst in the latter the influence of pH is only very slight. Thus PASTUSKA's experiments on silica gel G yielded the same results in the pH range of 5–10; the observed M_G values remained constant within the limits of experimental error ($\pm$ 0.02).

Normal paper electrophoresis apparatus was used by these authors, without cooling.

The times of electrophoresis were generally 90 min, whilst the same separations required longer periods when paper electrophoresis was used. Within certain limits, the duration of electrophoresis is not important: experiments in which the time of

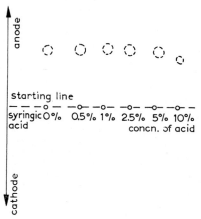

Fig. 3. Variation of the M_G value with the quantity of HCl added to the solution of syringic acid.

migration was varied between 60 and 150 min showed no change in the value of M_G.

The question of whether additions of acid to the starting mixture had any influence on M_G was also of interest, since examination of natural products frequently involves preliminary acid hydrolysis (e.g. the case of glucosides). Such acid additions influence the results in paper chromatography, and it may be very difficult to eliminate the acid from the solution. The results obtained on silica gel G with solutions of syringic acid containing up to 10% of hydrochloric acid are shown in Fig. 3. It can be seen that the mobility is influenced by acid concentrations in excess of 5%. The effect of acid concentration on M_G is slightly greater for certain phenols, e.g. pyrogallol, but concentrations of up to 1% can be safely used.

The results obtained for phenols by PASTUSKA AND TRINKS are listed in Table 1, which shows the M_G values for all the substances examined relative to m-hydroxybenzoic acid, the direction of migration, the colour produced on reaction with diazotised benzidine, and the quantity of sample used (expressed in µg). As stated above, the M_G values remain constant for quantities between 1 and 20 µg.

As can be seen from this table, the compounds examined tended to migrate further on kieselguhr G than on the silica gel. Thus for pyrocatechol, M_G was 0.85 on kieselguhr and 0.34 on silica gel, and for pyrogallol the corresponding values of M_G were 0.71 and 0.38. All the compounds migrated towards the anode on kieselguhr, whilst e.g. quinone and kojic acid were displaced towards the cathode on silica gel. The two values given for hydroquinone are due to the fact that on silica gel G the hydroquinone is largely oxidised to quinone, which migrates towards the cathode with $M_G = 0.06$. The value of $M_G = 0.42$ corresponds to the unoxidised portion of hydroquinone, which migrates anodically. Hydroquinone was completely oxidised to quinone on kieselguhr G. The displacement of methyl gentisite towards the cathode on silica gel may be ascribed to the formation of a quinone system. Free gentisic acid migrates towards the anode since it cannot assume a quinone structure owing to the possibility of hydrogen bonding between the carboxylic group and the neighbouring hydroxyl.

PASTUSKA AND TRINKS also revealed the existence of certain relationships between the structures and electrophoretic mobilities of the examined compounds. It was for example observed that on silica gel G, the esters of hydroxybenzoic acids possessed lower mobilities than the free acids themselves, and that the mobility was lowered by the presence of side chains (thus p-cumaric acid has a lower mobility than p-hydroxybenzoic acid, and caffeic acid than protocatechuic acid).

PASTUSKA AND TRINKS observed important differences in the behaviour of various o-diphenols during paper and thin-layer electrophoresis. According to MICHL[10] o-diphenols cannot be separated by paper electrophoresis using 0.1 N H_3BO_3 as the electrolyte, since they have approximately the same mobilities. As can however be seen from Table 1, such separations present no difficulties on a thin layer of kieselguhr G (pyrocatechol $M_G = 0.85$; ethyl protocatechuate $M_G = 0.67$; pyrogallol $M_G = 0.71$; ethyl gallate $M_G = 0.61$). This difference in the electrophoretic behaviour may be explained on the assumption that with both silica gel G and kieselguhr G, the boric acid in the thin layer forms complexes with the o-diphenols, and that these

References p. 68

TABLE 1

ELECTROPHORESIS OF PHENOLS

Substance	On Silica Gel G				On Kieselguhr G			
	M_G	Direction of migration	Coloration with diazotised benzidine	Quantity of substance (μg)	M_G	Direction of migration	Coloration with diazotized benzidine	Quantity of substance (μg)
Phenol	0.06	anode	yellow	10.00	0.06	anode	clear yellow	5.00
Pyrocatechol	0.34	anode	yellow	2.50	0.85	anode	yellow grey	2.50
Resorcinol	0.07	anode	red	1.25	0.04	anode	red	1.25
Hydroquinone	0.42	anode	yellow brown	10.00	0.05	anode	yellow	10.00
	0.06	cathode	yellow					
Quinone	0.04	cathode	yellow	10.00	0.05	anode	yellow	10.00
Phloroglucinol	0.06	anode	violet	1.25	0.04	anode	violet	1.25
Pyrogallol	0.38	anode	red brown	2.50	0.71	anode	brown violet	2.50
Salicylic aldehyde	0.04	anode	brown	—	0.00	anode	light brown	25.00
Salicylic acid	0.49	anode	+	25.00	0.99	anode	+ +	25.00
m-Hydroxybenzoic acid	1.00	anode	yellow	15.00	1.00	anode	yellow	15.00
p-Hydroxybenzoic acid	0.77	anode	yellow	10.00	0.86	anode	yellow	10.00
Gentisic acid	0.56	anode	light red	5.00	0.92	anode	yellow brown	5.00
Methyl gentisite	0.07	cathode	light red	5.00	0.05	anode	yellow brown	5.00
α-Resorcylic acid	0.64	anode	red brown	5.00	0.90	anode	yellow red	5.00
Protocatechuic acid	0.83	anode	clear brown	5.00	1.12	anode	yellow brown	5.00
Ethyl protocatechuate	0.28	anode	clear brown	5.00	0.67	anode	grey brown	5.00
Gallic acid	0.64	anode	orange	2.50	1.05	anode	orange	2.50
Ethyl gallate	0.42	anode	orange	5.00	0.61	anode	violet	5.00
Cinnamic acid	0.85	anode	clear brown	25.00	0.97	anode	+ +	25.00
p-Cumaric acid	0.72	anode	yellow brown	5.00	0.86	anode	yellow	5.00
Caffeic acid	0.66	anode	brown	2.50	1.07	anode	yellow brown	2.50

THIN-LAYER ELECTROPHORESIS

Guaiacol	0.06	anode	yellow brown	—	0.06	anode	yellow brown	—
Dimethyl ether of resorcinol	0.00	—	+	25.00	0.04	anode	orange	—
Anisic acid	0.95	anode	+	25.00	1.05	anode	+	25.00
o-Vanillin	0.05	anode	red	10.00	0.08	anode	yellow red	10.00
Isovanillin	0.04	anode	yellow	10.00	0.03	anode	red orange	10.00
Vanillin	0.04	anode	yellow	25.00	0.05	anode	yellow	25.00
Vanillic acid	0.86	anode	yellow brown	5.00	0.87	anode	brown	5.00
Veratric acid	0.82	anode	++	25.00	0.92	anode	++	25.00
Syringic aldehyde	0.06	anode	yellow brown	25.00	0.06	anode	light yellow	25.00
Syringic acid	0.74	anode	yellow red	5.00	0.83	anode	yellow red	5.00
Ferulic acid	0.73	anode	yellow brown	5.00	0.77	anode	brown	5.00
Isoferulic acid	0.76	anode	red	5.00	0.78	anode	red	5.00
Chlorogenic acid	0.35	anode	yellow brown	25.00	0.75	anode	yellow brown	25.00
Kojic acid	0.07	cathode	brown violet	25.00	0.07	anode	red	25.00
Umbelliferone	0.92	anode	blue (u.v.)	25.00	1.03	anode	blue (u.v.)	25.00
	0.55	anode	red orange		0.70	anode	red orange	
	0.04	cathode	blue (u.v.)		0.05	anode	blue (u.v.)	

+ = with SbCl₅. ++ = with alk. KMnO₄.

Naphthols

In subsequent work[4] in 1962, PASTUSKA AND TRINKS investigated the electrophoresis of a series of naphthols, using the same electrolyte as for the phenols, and employing the technique described above. The results obtained on thin layers of silica gel G are shown in Table 2.

TABLE 2
ELECTROPHORESIS OF NAPHTHOLS ON SILICA GEL G

Substance	Distance travelled (in mm)	Direction of migration	Coloration with diazotised benzidine
α-naphthol	18	cathode	grey brown
β-naphthol	12	cathode	violet
1,3-dihydroxynaphthalene (Merck)	15	cathode	dark brown-violet
	40 ⎫ weak	anode	yellow brown
	65 ⎭ spots	anode	yellow brown
4-nitrobenzene-(1-azo-4)-naphthol (magneson II)	10	cathode	light violet
sodium salt of α-naphthol-5-sulphonic acid	75	anode	clear red brown
1,8-dihydroxynaphthalene-3,6-disulphonic acid (chromotropic acid)	100	anode	clear red brown

Amines and amino acids

Thin-layer electrophoresis of amines and amino acids was also studied by PASTUSKA AND TRINKS and by HONEGGER. In their second investigation[4], PASTUSKA AND TRINKS studied the electrophoretic behaviour of these compounds on thin layers of silica gel G, prepared with borax, using a mixture of EtOH (80 ml), H_2O (30 ml), and sodium acetate (2 g) as the electrolyte. The pH was adjusted to 12 with NaOH. Variations in the pH are not important in this case, as pH values lying between 10 and 12 produce the same M_G values. Electrophoresis was carried out for 120 min, in a field of 10 V/cm. The spots were developed by treatment with ninhydrin and subsequent heating at 90–100°C. The M_G values obtained for the amines are shown in Table 3. Methylamine was used as the reference compound. Triethanolamine was developed with alkaline permanganate, since as a tertiary amine it does not react with ninhydrin. Table 4 shows the M_G values for the amino acids, with glycine as the reference compound. It should be noted that all the compounds examined migrated towards the cathode, and not towards the anode as might have been expected. This result may be explained by variation of the pH of the layer during electrophoresis.

TABLE 3
ELECTROPHORESIS OF AMINES ON SILICA GEL

Substance	M_G	Direction of migration	Coloration with ninhydrin
methylamine	1.00	cathode	red violet
ethylamine	0.69	cathode	red violet
n-butylamine	0.82	cathode	red violet
ethanolamine	1.09	cathode	red violet
ethylenediamine	0.32	cathode	red violet
triethanolamine	0.51	cathode	red violet

TABLE 4
ELECTROPHORESIS OF AMINO ACIDS ON SILICA GEL G

Substance	M_G	Direction of migration	Coloration with ninhydrin
glycine	1.00	cathode	orange
β-alanine	0.83	cathode	blue
DL-lysine monohydrochloride	0.81	cathode	red
L-(—)-cystine	0.0	—	red violet
DL-methionine	0.92	cathode	red
DL-phenylalanine	0.98	cathode	red brown
DL-tryptophan	0.90	cathode	blue
L-histidine monohydrochloride	0.46	cathode	red

The publication of PASTUSKA AND TRINKS' first paper in 1961 coincided with a paper by HONEGGER[5] describing the separation of amines and amino acids, both by thin-layer electrophoresis and by electrophoresis followed by chromatography on the same layer. The strips used by HONEGGER were of silica gel G, kieselguhr G, and alumina G. The experiments were carried out with apparatus similar to that described above, with cooling in the lower part of the support. The separations obtained and their reproducibilities are shown in Fig. 4.

Methylamine ($E_f = 1$) and cysteine ($E_f = 0$) were taken as the reference compounds. As can be seen from Fig. 4, a good separation can be obtained in one hour, the reproducibility of E_f values being 2–3%.

Fig. 5 illustrates the results obtained by HONEGGER using electrophoresis and subsequent chromatography on the same layer of silica gel G, with a mixture of amino acids and amines. The values obtained by the electrophoresis of a mixture of methylamine (M), cadaverine (C), mescaline (Mc), and cysteine (Cy) are given in the upper part of the figure for the sake of comparison. It is of interest to note that whereas a

References p. 68

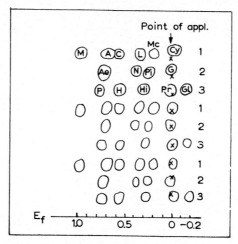

Fig. 4. Electrophoresis of three mixtures of amines and amino acids on a thin layer of silica gel G. Electrolyte: sodium citrate buffer (0.1 N), pH = 3.8; 440 V; 19.6 mA; 1 h. (1) M = methylamine; A = ethylamine; C = cadaverine; L = lysine; Mc = mescaline; Cy = cysteine. (2) Ae = ethanolamine; N = noradrenalin; Pi = piperidine; G = glycine. (3) P = propylenediamine; H = histamine; Hi = histidine; Pr = proline; Gl = glutamic acid.

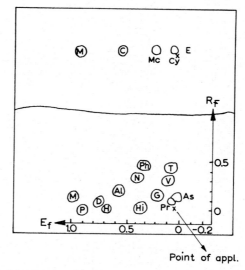

Fig. 5. Electrophoresis of a mixture of amines and amino acids followed by chromatography on the same thin layer of silica gel G. Electrolyte: 2 N acetic acid/2 N formic acid (1 : 1), pH = 2; 460 V; 12.6 mA; 1 h. Eluent for chromatography: n-butanol/acetic acid/H$_2$O (3 : 1 : 1).

separation of *e.g.* glycine and proline could not be obtained by electrophoresis alone, it could be achieved by the application of the combined technique.

Dyestuffs

PASTUSKA AND TRINKS[4] also carried out some investigations on azo dyes and on other dyestuffs of the phthalein group. The results listed in Table 5 were obtained

TABLE 5
ELECTROPHORESIS OF DYESTUFFS ON SILICA GEL

Substance	In the acidic electrolyte			In the alkaline electrolyte		
	Distance travelled (in mm)	Direction of migration	Colour of spots	Distance travelled (in mm)	Direction of migration	Colour of spots
Methyl orange	25	anode	yellow red	0 7	— anode	orange —
Methyl red	0	—	red	34	anode	yellow
Dimethyl yellow	0	—	yellow	0	—	yellow
Congo red	0	—	blue	30	anode	red
Eriochrome black T	0	—	black	0	—	red
Crystal Ponceau	65 80	anode anode	red red	35	anode	red
Phenolphthalein	14	cathode	(with NaOH) red	16	cathode	red
Thymolphthalein	0	—	uncoloured	0	—	blue
Bromophenol blue	51	anode	yellow	15	anode	blue
Fluorescein	28	anode	yellow green	22	anode	yellow
Rhodamine B	0 25 37 56 75	— anode anode anode anode	red visible under u.v. red and blue	0	—	red
Alizarin S	0 52	— anode	visible under u.v.	0	—	visible under u.v.
Neocarmine W	0 78 85	— anode anode	blue yellow red	0 28 38 54	— anode anode anode	blue yellow red red

on silica gel G using an acidic electrolyte (80 ml ethanol, 30 ml water, 4 g H_3BO_3, 2 g sodium acetate; pH 4.5) and an alkaline electrolyte (80 ml ethanol, 30 ml water, 2 g sodium acetate; pH adjusted to 12 with NaOH). The results were naturally different in the two cases. Thus methyl red and Congo red, for example, remained at the point of application in the acidic electrolyte whilst in the alkaline electrolyte they migrated over 34 and 30 mm respectively.

Separation and determination of iodates and periodates

Thin-layer electrophoresis has also been applied to the separation of inorganic ions[6], and particularly for the resolution of iodate–periodate mixtures[7] which, ac-

References p. 68

cording to many authors[11-15] cannot be quantitatively separated on paper. In paper electrophoresis the periodate ion gives rise to two or three spots, probably by the formation of complexes with the cellulose of the strip or possibly owing to partial reduction. Of the other supports tried, the best results were obtained with plaster of Paris. This adsorbent had previously given good results in a few chromatographic separations of inorganic ions[16]. Iodate and periodate could not however be separated by chromatography, since on elution with 0.05 M ammonium carbonate the periodate formed a single spot at the point of application whilst the iodate migrated with the front of the eluent and formed a long tail.

Preparation of the thin layers and the apparatus used[7] for the electrophoresis of iodate–periodate mixtures are similar to those described above. Complete separation of the periodate ions (which remain at the point of application) and the iodate ions (which migrate a few centimetres towards the anode) is achieved using 0.05 M $(NH_4)_2CO_3$ as the electrolyte and a potential of 300–400 V, over 90–120 min.

For quantitative determinations, portions of the plaster of Paris strip containing the iodate and the periodate respectively were placed in flasks and broken up, and the quantities of iodate and periodate present were determined by titration with 0.01 N thiosulphate in the presence of excess KI, in acid solution. To check that the mixture used for the separation was stable over an appreciable period of time, the periodate content was checked before each separation by titration with 0.01 N arsenite in alkaline solution, whilst the sum of iodate and periodate was determined by titration with 0.01 N thiosulphate in acid solution. The results obtained in the separation of various quantities of potassium iodate and periodate are shown in Table 6. The values in the 1st and 3rd column of the table refer to quantities of iodate and periodate present in the mixture, determined as described above. The values in

TABLE 6

QUANTITATIVE ANALYSIS OF MACRO AMOUNTS OF PERIODATE AND IODATE

Periodate present (mg)	Periodate found (mg)	Iodate present (mg)	Iodate found (mg)
0.30	0.30	1.10	1.14
0.86	0.86	1.58	1.56
0.36	0.36	0.27	0.27
1.06	1.08	2.48	2.54
0.47	0.47	0.69	0.65
0.19	0.19	1.36	1.36
1.06	1.12	2.48	2.40
0.47	0.48	0.67	0.67
0.21	0.21	1.26	1.24
0.12	0.12	0.23	0.24
0.36	0.36	0.25	0.25
0.30	0.30	1.10	1.13

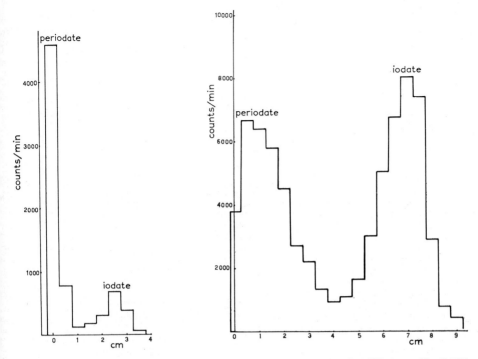

Fig. 6. Electrophoretic separation on plaster of Paris of a radioactive $K^{128}IO_4/K^{128}IO_3$ mixture obtained by irradiating KIO_4 and heating at 240°C.

Fig. 7. As in Fig. 6, using Whatman No. 3 MM paper in place of the plaster of Paris for electrophoresis.

the 2nd and 4th column are the values found after the separation. The results are clearly in good agreement in the twelve separations reported.

These authors also studied the question of whether this electrophoretic technique could be applied to the separation of products obtained from the Szilard–Chalmers reaction, exposing potassium periodate to a neutron flux of 10^{12} cm^{-2} · sec^{-1}. After irradiation the samples were heated at 240°C for 15 min. The activity of the ^{128}I produced was determined directly on thin layers using the apparatus described by LINSKENS[17], and the electrophoresis was carried out under the conditions described above. The results are shown in Figs. 6 and 7. As can be seen from Fig. 6, good separation is obtained on plaster of Paris, and the quantity of periodate formed under the conditions described is about 80%, whilst the iodate represents about 20%. The above figures correspond to those of ATEN et al.[18] who arrived at the result by another method. The periodate peak obtained by paper electrophoresis (Fig. 7) under the same conditions as those of Fig. 6 is less clearly defined, and the activity due to periodate is only 40%. Comparison of these results suggests that under the usual conditions of paper electrophoresis, considerable quantities of periodate are reduced to iodate.

References p. 68

5. CONCLUSIONS

Thin-layer electrophoresis can compete with paper electrophoresis as a technique, since it is capable of the same reproducibility, the time required for the separation is generally shorter, and the pH of the electrolyte may vary within certain limits without affecting the final results. Furthermore, certain systems such as o-diphenols, and iodate–periodate mixtures, which could not be separated with the aid of paper electrophoresis, can be resolved by electrophoresis on thin layers.

REFERENCES

1. R. Consden, A. H. Gordon and A. J. P. Martin, *Biochem. J.*, 40 (1946) 33.
2. E. Stahl, *Pharmazie*, 11 (1956) 633; *Chemiker Ztg.*, 82 (1958) 323.
3. G. Pastuska and H. Trinks, *Chemiker Ztg.*, 85 (1961) 535.
4. G. Pastuska and H. Trinks, *Chemiker Ztg.*, 86 (1962) 135.
5. C. G. Honegger, *Helv. Chim. Acta*, 44 (1961) 173.
6. B. Pfrunder, R. Zurflüh, H. Seiler and H. Erlenmeyer, *Helv. Chim. Acta*, 45 (1962) 1153.
7. F. Dobici and G. Grassini, *J. Chromatog.*, 10 (1963) 98.
8. Th. Wieland and G. Pfleiderer, *Angew. Chem.*, 67 (1955) 257.
9. G. Marten, *Pharmazie*, 10 (1955) 602.
10. H. Michl, *Monatsh.*, 83 (1952) 737.
11. M. Lederer, *Anal. Chim. Acta*, 17 (1957) 606.
12. J. Jach, H. Kawahara and G. Harbottle, *J. Chromatog.*, 1 (1958) 501.
13. D. Gross, *Chem. Ind. (London)*, (1957) 1597.
14. G. Grassini and M. Lederer, *J. Chromatog.*, 2 (1959) 326.
15. G. B. Belling and R. E. Underdown, *Anal. Chim. Acta*, 22 (1960) 203.
16. B. N. Sen, *Anal. Chim. Acta*, 23 (1960) 152.
17. H. F. Linskens, *J. Chromatog.*, 1 (1958) 471.
18. A. H. W. Aten, Jr., G. K. Koch, G. A. Wesselink and A. M. de Roos, *J. Am. Chem. Soc.*, 79 (1957) 63.

SPECTROMÉTRIE DE MASSE ET CHROMATOGRAPHIE EN COUCHE MINCE

M. FÉTIZON

Institut de Chimie des Substances Naturelles, Gif-sur-Yvette, Seine-et-Oise (France)

1. INTRODUCTION

La synthèse totale d'une substance organique complexe, telle que la plupart des substances naturelles, s'effectue habituellement en de nombreuses étapes, de telle sorte qu'il est nécessaire, soit de partir de quantités relativement importantes de matière première, en utilisant les méthodes classiques de purification et d'analyse, soit d'adapter ces méthodes à une échelle très inférieure à celle couramment en usage.

Ces dernières années, deux techniques nouvelles sont apparues, dont la combinaison est susceptible de rendre les plus grands services: la séparation par chromatographie sur couche mince d'adsorbant, et la spectrométrie de masse. La chromatographie préparative sur couche mince permet de travailler sans difficulté sur dix ou vingt milligrammes de substance. Le constituant intéressant peut être isolé, et sa pureté contrôlée, par exemple par chromatographie analytique selon STAHL.

Dix milligrammes de cristaux ne constituent jamais une quantité massive, mais c'est parfois le stock mondial: il importe donc d'en dilapider le minimum.

Certes, dix milligrammes suffisent à déterminer un spectre infra-rouge, une courbe de dispersion rotatoire ou de dichroïsme circulaire. Cependant, il est nécessaire de purifier de nouveau pour la microanalyse, qui devient souvent à peu près irréalisable.

Le spectromètre de masse est alors l'outil le mieux adapté à cette situation car, si peu qu'il reste de l'échantillon pur, c'est suffisant pour cet énorme appareil au si minuscule appétit. On en déduit, non seulement la masse moléculaire, mais souvent la formule brute, et même des informations structurales du plus haut intérêt.

2. DISCUSSION

Il n'est guère indispensable de rappeler le principe de la chromatographie préparative sur couche mince d'alumine ou de silice.

L'alumine rendue fluorescente avec du morin[1] est d'un emploi particulièrement aisé[2]. L'épaisseur de la couche utilisée doit toutefois demeurer assez faible, de l'ordre de 1 m/m au plus, pour que les séparations soient satisfaisantes. Après séchage, on repère la position des zones adsorbées, sous lumière ultraviolette, et aspire directement dans un ballon ou une petite colonne à chromatographie la région intéressante. Il reste à éluer avec un solvant convenable, et, éventuellement, à recristalliser le produit obtenu.

Entre le moment où on dispose du produit brut, et celui où la substance pure est isolée, compte tenu de la préparation de la plaque, il s'écoule en moyenne quarante cinq minutes.

Il est même souvent plus rapide de faire plusieurs chromatoplaques en parallèle qu'une seule chromatographie classique.

Il est probablement plus utile d'exposer en quoi consiste la spectrométrie de masse, ce qu'on peut en attendre, et comment son emploi judicieux, en liaison étroite avec les techniques chromatographiques peut se révéler efficace.

Soit donc une molécule A : B, à l'état vapeur à très faible pression, soumise à un bombardement d'électrons d'une énergie de 10 électron-volts environ (pouvant si on le désire, s'élever jusqu'à 70 électron-volts).

Dans ces conditions, la molécule perd un électron.

$$A : B + e^- \rightarrow 2e^- + (A \cdot B)^{\oplus}$$

L'espèce $(A \cdot B)^{\oplus}$, si elle est suffisamment stable, vit assez longtemps pour être accélérée par un champ électrique, déviée par un champ magnétique, de telle sorte que l'ion en se déchargeant fournisse un signal caractéristique du rapport de sa masse m à sa charge e.

Pratiquement, on obtient sur l'enregistreur un pic à $m/e = M$ ou M est la masse moléculaire de A:B.

La plupart des appareils modernes distinguent aisément M et $M+1$ quand M est de l'ordre de 300 ou 400, et souvent même 500 à 600.

La première information qu'on va donc tirer du spectre de masse de A:B, c'est, dans beaucoup de cas, sa masse moléculaire. C'est en général suffisant pour affirmer qu'une telle réaction a effectivement fourni le résultat souhaité.

On peut toutefois, obtenir davantage de précisions.

En effet, si A:B est une molécule organique, elle contient une certaine proportion de carbone-^{13}C, d'oxygène-^{18}O, d'azote-^{15}N, etc. A côté du pic moléculaire M, on observe donc des pics satellites à $M+1$, $M+2$, etc., d'intensité très rapidement décroissante, parce que la probabilité pour qu'une molécule possède deux carbones-^{13}C par exemple (pic $M+2$) est beaucoup plus faible que la probabilité pour qu'elle en possède un seul.

Néanmoins, et à condition de négliger d'autres effets, on peut calculer la hauteur des pics $M+1$ et $M+2$ en fonction du pic M pour diverses structures de même poids moléculaire[3,4].

Par exemple, les substances de formules brutes $C_{10}H_{12}$ et $C_8H_8N_2$ ont même masse moléculaire observable ($M = 132$). Cependant, on trouve[3]:

	M	$(M+1)/M$	$(M+2)/M$
$C_{10}H_{12}$	132	11,0 %	0,6 %
$C_8H_8N_2$	132	9,52 %	0,31 %

Le simple examen du pic moléculaire et des pics satellites permet donc très souvent de trancher entre deux formules brutes possibles, sans pour cela faire appel à des appareils à haute résolution, encore peu répandus.

Déterminer la masse moléculaire, et éventuellement la formule brute sur une fraction de milligramme, est de toute première importance, mais on peut glaner bien d'autres renseignements encore.

Revenons à la molécule $(A \cdot B)^{\oplus}$ ionisée. La liaison à un seul électron peut se rompre, soit en ion $B^{\oplus}$ et radical $A \cdot$, soit en molécule neutre A et ion radical $\cdot B^{\oplus}$.

Ces ruptures, ou fragmentations, peuvent d'ailleurs être de deux types: les ruptures simples, et celles qui s'accompagnent de transferts d'hydrogène.

Il n'est guère possible d'entrer dans le détail de ces fragmentations dont l'étude constitue d'ailleurs un domaine en évolution très rapide. On va donc se borner à quelques indications.

Tout d'abord, une remarque est essentielle: pour autant qu'on le sache, tout ce qui stabilise au sens classique du terme, l'ion $B^{\oplus}$, le radical $A \cdot$ ou le fragment A facilite la rupture de A:B en A et $B^{\oplus}$, $B^{\oplus}$ étant d'ailleurs seul observé (les molécules neutres ne sont évidemment pas accélérées par le champ électrique).

C'est pourquoi, on observe la coupure d'une chaîne de préférence à une ramification (un ion $C^{\oplus}$ tertiaire est beaucoup plus stable qu'un carbocation secondaire ou surtout primaire).

La conjugaison peut également grandement faciliter la formation de l'ion positif. C'est ainsi que, par exemple, un diterpène tel que (I) fournit à m/e : 81 un pic très intense (en général, on observe une coupure benzylique).

De même, l'amine (II) donne un très grand pic à m/e : 158, caractéristique du groupe diméthylaminoalkyl.

Ici, les électrons du doublet de l'azote stabilisent le cation.

Une fragmentation d'un grand intérêt dans les cycles hexaatomiques a été indiquée par BIEMANN: c'est la coupure rétro Diels–Alder[5,6]. Ainsi:

La position de la double liaison est d'une grande importance:

pas de fragmentations rétro Diels-Alder

Les fragmentations avec réarrangement d'hydrogène sont bien plus variées, e souvent encore mal connues. Elles s'effectuent le plus souvent par l'intermédiair d'états de transition à six centres ou à quatre centres:

$m/e : 74$

$m/e : 124$

3. APPLICATION

Quelques exemples suffiront pour illustrer l'application de la spectrométrie de mass à l'analyse de substances isolées en petites quantités par chromatographie sur couch mince.

Récemment, une corrélation a été établie entre les stéroïdes, dont la stéréochimie est parfaitement connue, et les diterpènes du groupe pimarique, où la situation était moins claire.

A cet effet, le sandaracopimaradiène (III) a été préparé[7] en de nombreuses étapes à partir d'hydroxy-3β-androstène-5-one-17 (IV).

Il est certain que l'emploi des techniques usuelles eût conduit à effectuer la dégradation en partant de quantités prohibitives d'une matière première relativement coûteuse.

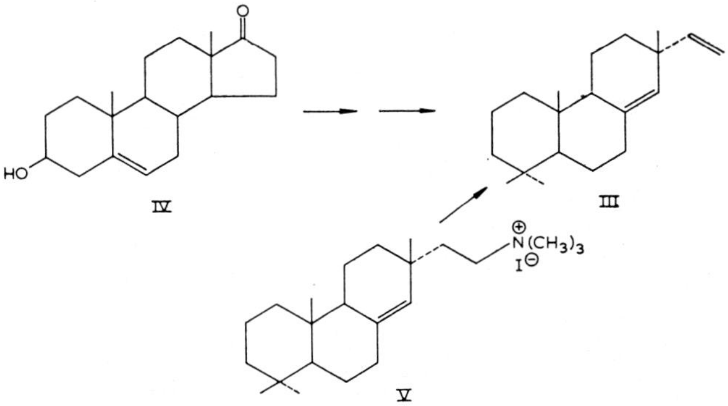

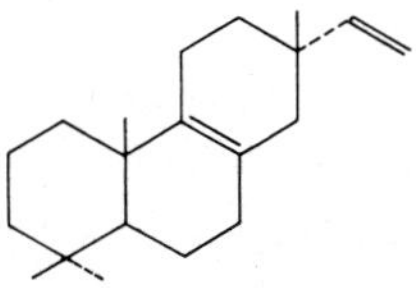

La dernière étape par exemple, réalisée sur 44 mg de (V), a fourni 19 mg du produit attendu, cristallisé malgré son bas point de fusion (F: 41°) sur lequel on a déterminé le spectre infra-rouge, la dispersion rotatoire, et le spectre de masse. Celui-ci a confirmé la masse moléculaire, et la position d'une des liaisons éthyléniques. En effet, un pic intense à m/e : 137 s'interprète de la manière suivante:

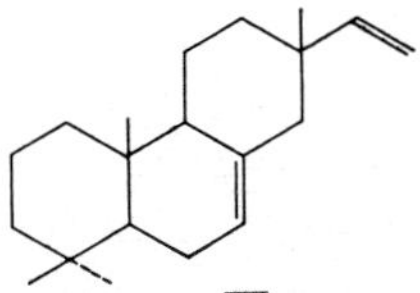

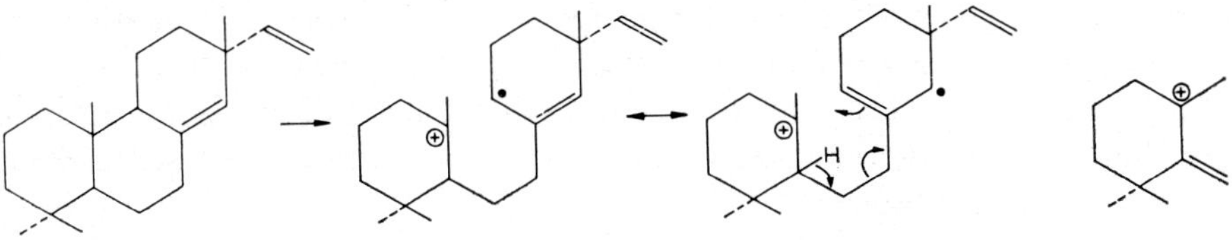

m/e : 137

Ce pic n'existe plus dans le cas de l'isomère (VI), ni dans l'isopimaradiène (VII), ni dans le rimuène, qui ne possède certainement pas de liaison éthylénique $\triangle^{8(14)}$.

Le sel d'ammonium quaternaire (VIII), traité par le butyllithium comme (V), fournit une substance (IX) qui est un alcool[8].

Le spectre de masse comporte un pic à m/e : 333 et un autre à m/e : 330. Comme il ne peut exister de pic à $M-3$, il faut interpréter ce spectre en prenant m/e : 330

Bibliographie p. 74

comme $M-18$ (—H_2O) et m/e : 333 comme $M-15$ (perte d'un méthyle). La chaîne butyle a donc été greffée quelque part sur la molécule initiale. L'absence de pic moléculaire est d'ailleurs un phénomène général pour les alcools tertiaires.

Les pics les plus importants à m/e : 192, m/e : 177 et m/e : 138 peuvent s'interpréter soit avec la structure (IXa), soit avec la structure (IXb).

La dégradation de très petites quantités de l'alcool (IX) a permis d'obtenir la lactone (X), identifiée après séparation sur plaque par son spectre infra-rouge (γ-lactone), son point de fusion et son spectre de masse (pic moléculaire). La structure de l'alcool est donc (IXb). Dans ce cas, la chimie classique, la spectrométrie de masse et la séparation par chromatoplaque ont collaboré pour donner la solution à un problème ou chaque méthode, prise séparément, eût échoué.

Ainsi donc, la spectrométrie de masse, jointe à la séparation par chromatoplaque, permet d'obtenir des résultats qui eussent été impensables, ou tout au moins, difficilement accessibles voici quelques années.

Il existe hélas, un envers à cette médaille: si chacun peut acquérir le matériel nécessaire aux chromatographies, le prix de revient de l'heure d'utilisation d'un spectrographe de masse demeure prohibitif, sauf peut-être pour un Institut de Recherches de dimensions convenables. Il semble qu'il en soit de la chimie organique d'aujourd'hui comme de la physique d'il y a vingt ans: ces gros appareils coûteux sonnent le glas de l'artisanat.

BIBLIOGRAPHIE

1. H. Brockmann et F. Volpers, *Chem. Ber.*, 80 (1947) 77.
2. V. Černy, J. Joska et L. Lábler, *Collection Czech. Chem. Commun.*, 26 (1961) 1658.
3. K. Biemann, *Mass Spectrometry. Organic Chemical Applications*, McGraw Hill, New York, 1962.
4. F. W. Lampe et F. H. Field, *Tetrahedron*, 7 (1959) 189.
5. K. Biemann, *Angew. Chem.*, 74 (1962) 102.
6. H. Audier, M. Fétizon et W. Vetter, *Bull. Soc. Chim. France*, (1963) 1971.
7. M. Fétizon et M. Golfier, *Bull. Soc. Chim. France*, (1963) 167.
8. H. Audier, S. Bory et M. Fétizon, *Bull. Soc. Chim. France*, sous presse.

THIN-LAYER CHROMATOGRAPHY OF STEROIDS

R. NEHER

Pharmaceutical Research Department, CIBA Ltd., Basel (Switzerland)

1. INTRODUCTION

We have all heard so much about thin-layer chromatography (TLC) in general that a discussion of the application of this method to steroids can be confined to a few particular points, a few examples, and a few critical remarks.

There is little point in making a distinction between adsorption and partition processes here, as all possible gradations between the two can occur, depending on the solvent system used. The TLC of lipophilic steroids is usually a question of adsorption or reversed-phase partition, while for that of the more hydrophilic steroids the normal partition process is also used. Amounts of 0.01–100 μg of the steroids may be employed for this purpose, depending on the detecting sensitivity and solubility in the systems.

As adsorbents or as supporting media silicic acid (silica gel), alumina, and cellulose have proved most useful, although layers of calcium sulphate or kieselguhr are also suitable in special cases. For most purposes silica gel, with or without a binder, is quite adequate. I shall not add any comments here on the quality or pre-treatment of adsorbents, on thin-layer preparation, or on the equilibration and saturation of chromatography jars, since these points have been discussed at some length earlier on in this meeting.

However, what I should like to discuss in greater detail is the choice of solvents and detection methods.

2. SOLVENTS

For adsorption and partition type TLC, the choice of solvents depends on the polarity of the steroids to be chromatographed, according to the eluotropic and mixotropic series. Table 1 gives the mixotropic series of solvents, according to HECKER[1], which is very similar to the eluotropic series, beginning with weakly polar solvents. The further from each other two solvents are in this series, the less they are miscible with each other, and thus the greater the number of compounds which can be simultaneously separated using these solvents, as long as the compounds to be separated are soluble enough.

A very rough indication can be obtained with the aid of the micro-circular method of STAHL[2] as discussed in his paper earlier in this Volume (pp. 1ff).

TABLE 1
THE MIXOTROPIC SERIES OF SOLVENTS, ACCORDING TO HECKER

paraffin oil	pyridine and homologs
hexane	tetrahydrofuran
heptane	dioxan
iso-octane	acetone
cyclohexane	aniline
cyclopentane	phenol
carbon disulfide	ethanol and homologs
carbon tetrachloride	glycol monomethyl ether
toluene	methanol
benzene	acetic acid and homologs
ethylene trichloride	acetonitrile
ethylene dichloride	nitromethane
chloroform	formic acid
ethylene tetrachloride	morpholine
methylene chloride	formamide
diethyl ether and homologs	water
methyl ethyl ketone	inorganic acids
ethyl acetate and homologs	salt or buffer solutions

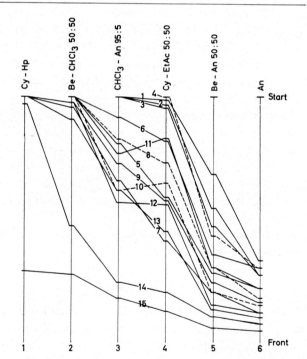

Fig. 1. Running distance of sterols and steroids of different polarity in solvent systems of different eluotropic properties. The numbers 1 to 6 (down) correspond to the numbered circles in Fig. 2. The R_F values in system No. 1 represent the mean values of the very similar values in cyclohexane (Cy) and heptane (Hp); Be = benzene, An = acetone, EtAc = ethyl acetate. The numbers 1–15 refering to joint R_F lines represent the steroids as given in the text.

T.L.C. OF STEROIDS

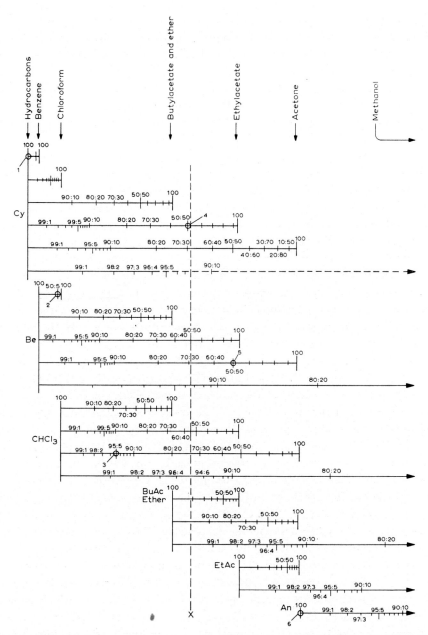

Fig. 2. Solvents and binary solvent mixtures arranged in the "equieluotropic series"[3] on the basis of *average* R_F values for 20 steroids. Each vertical line connects solvent mixtures of the same average eluotropic properties; pure methanol is somewhere out of the figure to the right. The first line section from the left or right in the "mixing line" joining two solvents always represents a mixture of 90 % and 99 % respectively of the nearest pure solvent, and 10 % or 1 % of the other, etc.

References p. 86

Fig. 1 gives a guide to the choice of solvents over a wide range of polarity. Here, the R_F-values of a number of sterols and steroids, *i.e.* from tetrahydrocortisol (No. 1) down to cholestane (No. 15), in six pure and mixed solvent systems of increasing eluotropic power are indicated graphically. The intermediary substances are: (2) cortisol, (3) cortisone, (4) corticosterone, (5) oestradiol-17β, (6) 5β-pregnane-3α, 20α-diol, (7) oestrone, (8) testosterone, (9) 3β-hydroxy-pregn-5-en-20-one, (10) androst-4-ene-3,17-dione, (11) cortexone, (12) progesterone, (13) cholesterol, and (14) cholesterol acetate.

Between solvent systems 1 and 6 there are, of course, hundreds of possible and useful mixtures for all types of neutral, phenolic, and weakly basic steroids; for bile acids, it is advisable to add an acid, so as to avoid tailing due to dissociation.

A finer eluotropic gradation of some suitable solvents and their binary mixtures on silica-gel layers for sterols and steroids of various polarity is given in Fig. 2. This series is based on mean R_F-values for the TLC of 20 steroids and a number of lipophilic dyes. On the left side of each mixing line is the 100% pure solvent of weaker polarity and on the right the 100% pure solvent of stronger polarity, the proportions in between being given on a logarithmic scale; pure methanol is somewhere out of the picture to the right. The pure solvents and their binary mixtures, *i.e.* the mixing lines, are now placed empirically under each other in such a way that the solvents of equal *average* eluting power for the tested steroids are to be found in the vertical dimension. Thus, the eluting power increases from left to right, whereas it remains practically constant along the vertical lines. The numbered circles 1–6 (from left to right) refer to the composition of the solvents of increasing polarity as shown in Fig. 1. On the other hand, the vertical line X has been drawn through the binary mixing lines in an arbitrary position. If, for example, three steroids of different polarity, such as testosterone, oestradiol-17β, and cholesterol were chromatographed together in the solvents whose composition is given by this line X, the R_F-values of each steroid would remain more or less constant in the various solvent systems.

The results in practice are shown in Fig. 3. At the top are listed the solvent systems composed according to the line X of Fig. 2. On the lines between start and front the experimental R_F-values of (1) testosterone, (2) oestradiol-17β, and (3) cholesterol have been marked. It may be seen that all R_F-values fluctuate about a certain mean value.

As you may have noticed from the preceding figures, and as far as polarity is concerned, not many solvents or mixtures are needed to cover the whole range of steroid chromatography; but in order to *separate* all the mixtures which are encountered, it is advisable to use several solvent systems of approximately equivalent eluting power, as may be seen from the zig-zag lines in Fig. 3, which intersect here and there; this indeed, is only to be expected, bearing in mind all the likely interactions between solute, adsorbent, and solvent molecules. This means that, if a steroid mixture cannot be separated in a solvent system of suitable polarity, an "equi-eluotropic" solvent system, but of completely different composition, may do the trick. This system of equi-eluotropic solvent mixtures is naturally also applicable to column chromatography.

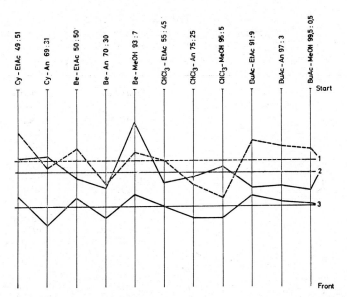

Fig. 3. R_F values of testosterone (1), oestradiol-17β (2) and cholesterol (3) in the binary solvent mixtures given by the vertical line X in Fig. 2. The R_F values of each steroid vary within a certain "equieluotropic" range about the mean value (straight lines). An = acetone, Be = benzene, BuAc = n-butyl acetate, Cy = cyclohexane, EtAc = ethyl acetate, MeOH = methanol.

For the solvent systems given here, very suitable test dyes are two lipophilic azo dyes: the weakly polar 4-amino-4'-nitro-azo-benzene (orange, = Oracetorange 2 R, CIBA) and the more strongly polar 4-nitro-2'-methyl-4'-diethanolaminoazobenzene (red, = Oracetred 2 G, CIBA).

I should also like to mention that most of the R_F-values given here are increased if alumina of medium-low activity is used instead of silica gel.

For partition type TLC of steroids, cellulose is by far the best carrier using the same one-phase or two-phase mixtures which have proved their worth for paper chromatography. If a two-phase system is used, it is often necessary to impregnate the thin layer with a spray or by dipping, e.g. with Zaffaroni-type systems. Some experiments with polar corticosteroids on kieselguhr-gypsum gave the same running rates as on cellulose using Zaffaroni-type systems. On the other hand, Bush-type solvent systems proved to be much more delicate when used on thin layers. For the reversed-phase technique, paraffin oil, silicone oil, undecane, etc. have been used as the stationary phase and acetic acid or acetonitrile–ketones as the mobile phase[5]. Other possibilities may be taken from the examples of reversed-phase methods as known in column and paper chromatography. Thus, the range of useful solvent systems for steroid TLC is enormous; the main problem in fact, lies in making an appropriate choice.

According to HEŘMANEK and co-workers[6], the influence which single substituents at distinct positions of the steroid nucleus exert on the R_F-value on alumina layers may be seen from Table 2. However, the magnitude of this influence is dependent

References p. 86

TABLE 2

INFLUENCE OF SINGLE SUBSTITUENTS AT DISTINCT POSITIONS OF THE STEROID NUCLEUS ON THE R_F VALUE ON ALUMINA LAYERS[6]

Be = benzene, EtOH = ethanol, Pe = petroleum ether

Steroid nucleus	Substituent X	R_F value Be/EtOH 98 : 2	
	$COOCH_3$	0.63	
	$OCOC_6H_5$	0.56	
	CN	0.50	
	$COCH_3$	0.51	
	$OCOCH_3$	0.42	
	=O	0.31	
	—OH	0.16	

	Substituent X	R_F value Pe/Be 100 : 0	50 : 50
	H	0.95	
	Cl	0.80	
	OCH_3	0.17	0.76
	$OCOCH_3$	0.13	0.73
	OH		0.09
	$N(CH_3)_2$	0.02	0.04

not only on the different positions and conformations (equatorial, axial), but to an even greater extent on the possible interactions of two or more substituents in the same molecule. Therefore, such series of adsorptivity are of rather limited practical value.

3. DETECTION METHODS

As far as detection and evaluation methods are concerned, practically the same considerations and methods apply here as in the well known case of paper chromatography though with some important exceptions.

On the one hand, inorganic chromatoplates (silica, alumina) offer the great advantage of being able to stand up to treatment with aggressive reagents, so that even very unreactive molecules can be detected by sensitive fluorescence methods or by carbonisation; on the other hand, so far it has unfortunately not proved possible to make direct use of two methods which are very suitable for steroids: the UV print and alkali fluorescence typical for $\triangle^4$-3-oxosteroids. The first method cannot be employed because of the strong absorption at 260 mμ by layers of SiO_2 or Al_2O_3; both methods are, however, possible with thin layers of cellulose.

The detection reagents which are most useful for steroid chromatography on chromatoplates and are usually applied as a spray are shown[5] in Table 3.

With many of these detecting reagents a quantitative evaluation, either directly on the plate or preferably in the eluate, is feasible under appropriate conditions.

It should be mentioned that in particular the various fluorescence reactions which

TABLE 3
DETECTING REAGENTS FOR TLC OF STEROIDS

1	H_2SO_4	conc. or 50% aqueous, for most of the steroids various colors and/or fluorescences (365 mμ light) of low specificity and high sensitivity (limit about 0.005 μg); charring; 50% in EtOH, same sensitivity but spots remain sharp
2	H_2SO_4-CH_3COOH 1 : 1,	red spots for cholesterol and its esters 2–4 μg; bile acids
3	Chlorosulfonic acid-CH_3COOH 1 : 2,	green spots with blue-violet fluorescence for cardenolides; sensitivity for cholesterol 0.025 μg
4	p-Toluenesulfonic acid 20% in EtOH,	fluorescing spots
5	H_3PO_4 20–70% in water or EtOH,	various colors and or fluorescences for many steroids, also with an additional spray of phosphomolybdic acid (blue spots)
6	Acids-aldehydes:	anisaldehyde/H_2SO_4/CH_3COOH, various colors and fluorescences, high sensitivity, low specificity anisaldehyde/perchloric acid, digitalisglycosides 0.02–0.1 μg vanillin/H_3PO_4 or H_2SO_4/EtOH
7a	Perchloric acid 20% in water,	} color and fluorescence
b	Trichloroacetic acid 25% in $CHCl_3$,	} reactions
c	Trichloroacetic acid-chloramine, glycosides, 0.01 μg UV 1 Vol. 25% in EtOH + 4 Vol. 3% in water	
8	Phosphomolybdic acid 10% in EtOH,	blue spots; cholesterol esters, <0.5 μg sensitivity depending on carrier and binder
9	Phosphotungstic acid 10% in EtOH,	red spots; cholesterol and esters
10	$SbCl_3$ in $CHCl_3$,	orange to violet spots for 3β-hydroxy-$\triangle^5$-steroids, 17α-hydroxy-$C_{18/19}$ steroids, bluish for 7-oxygenated $\triangle^5$-steroids; various colors and fluorescences for many steroids and related compounds
11	$SbCl_5$ in $CHCl_3$ or CCl_4	spots of less differentiated colors
12	2,4-Dinitrophenylhydrazine in MeOH/HCl, oxo-steroids, sensitivity depends on position of oxo group	
13	Isonicotinic acid hydrazide in EtOH/CH_3COOH, oxo-steroids	
14	m-Dinitrobenzene in MeOH/KOH,	17-oxo-steroids, weaker than on paper; modification with a first alkaline spray
15	Triphenyltetrazoliumchloride in MeOH/NaOH, reducing steroids, red spots	
16	Blue tetrazolium in water/NaOH, reducing steroids, blue spots, weaker than on paper (combination of yellow alkali fluorescence for $\triangle^4$-3-oxo-steroids only possible on cellulose layers!)	
17	Tollens reagent (silver diammine), for reducing substances, weaker than on paper	
18	2,4,2′,4′-Tetranitrodiphenyl 5% in benzene, cardenolides	
19	$FeCl_3$-$K_3Fe(CN)_6$ in water, or MeOH/water	blue spots for phenolic or enolic steroids, increase of color after HCl-spray
20	Diazonium salts from sulfanilic acid, (Pauly's reagent)	yellow to red spots for phenolic substances
21	Folin-Ciocalteu: NH_3 vapour, then reagent, blue spots with phenolic substances, α-ketols, α,β-diketones	
22	Boute reagent (NH_3 and NO_2 vapours),	yellow nitroso complexes with oestrogens
23	Iodine plateate in HCl,	steroid alkaloids and amino steroids; iodine bismuthate (Dragendorff reagent) is of lower sensitivity
24	$KMnO_4$-H_2SO_4, white spots on pink background for all oxidizable substances	
25	Rhodamin B 0.5% in EtOH, cholesterol esters, lipids	

can be obtained using sulphuric acid are often very sensitive (detection limit 0.01 μg and less), although this sensitivity varies *widely* from substance to substance within the same class. Therefore, too exclusive reliance on the fluorescence intensities of the various spots on the unknown chromatoplate may give a completely false picture of the relative concentrations; attention must also be paid to the evaluation of the purity of a substance to be analysed after one has found impurities which give 100 times more fluorescence per unit weight than the main substance. One may in fact occasionally be tempted to conclude that it is most difficult to find "pure" substances. It is, however, usually sufficient for the purposes of steroid chemistry if the substances are 99–99.9% pure with the exception of metallic impurities. A carbonisation reaction on the plate, combined with a photocopy, will thus give a more realistic result.

Another possibility is the use of fluorescent layers, with the various UV absorbing substances showing up as dark spots. This method has the advantage of not causing a chemical change in the substances in question. Fluorescent layers are obtained by adding various inorganic phosphors or organic fluorescent materials to the adsorbent or carrier such as morin, fluorescein, quinine sulphate, sulphosalicylic acid, etc.; with most of the commercially available inorganic phosphor layers, dark blue spots on a light green background are then produced with, for example, phenolic steroids, viewed in 360 mμ light, or $\triangle^4$-3-oxosteroids viewed in 250 mμ light; they can easily be observed when present in amounts down to 0.1 μg; sensitivity largely depends, however, on the quantity of phosphor and on the quality of the UV lamp used.

A further variation, in combination with a chemical reaction, is the fluorescein–bromine test for unsaturated compounds. On the other hand, the use of iodine vapour enables the substances to be made temporarily visible by means of their labile addition compounds with iodine, after which they can be eluted in their original form. Such a non-destructive method for zone development is particularly valuable for the preparative TLC and can also be achieved with lipophilic substances by spraying the TL chromatogram with water alone; zones of differing translucency are then obtained[5].

It is also possible to carry out several colour reactions on the same chromatoplate: this can be done, for example, by lightly spraying the first time and then, after the reaction is completed, by removing a layer of 50–100 μ; such a layer can easily be removed by sticking adhesive tape to the surface of the layer and then removing it again. This procedure can be repeated more than once.

A new detection method recently described for purines may also be of interest for steroids[7]. This method is based on microsublimation of the chromatographed substances from the chromatoplate to a clean, cooled glass plate a short distance away. Under suitable conditions, the sublimate is obtained in the form of a mirror image of the original chromatoplate, on which any desired detecting reaction can be carried out with a few μg, e.g. UV photoprints, microreactions, microcrystallisation and micromelting point.

Microreactions can also be carried out with a few μg directly on the chromatoplate, e.g. dehydration with sulphuric acid, oxidation with CrO_3 in acetic acid, reduction with 10% aluminium isopropoxide in benzene or with 0.5% KBH_4 in aqueous meth anol, bromination, hydrogenation, hydrolysis and esterification, and formation of nitroso complexes or of Girard derivatives with oestrogens.

After this rather exhausting enumeration of a multitude of possibilities at our disposal, let us consider two instructive examples which have been published recently.

4. EXAMPLES

The first example (Figs. 4 and 5) shows a system for the separation and characterisation of 24 oestrogens on thin silica gel G layers by LISBOA AND DICZFALUSY[8]; it consists of a combination of one- and two-dimensional TLC with some chemical reactions for derivative formation.

Chromatography in system A, *i.e.* EtAc–Cy–EtOH (45 : 45 : 10), separates 5 polar oestrogens from 19 less polar ones. The 5 polar oestrogens are then transferred to another plate and chromatographed bidimensionally in system B, *i.e.* EtAc/water/satd. Hx–EtOH (80 : 15 : 5) where complete separation is obtained. The less polar oestrogens are rechromatographed bidimensionally in system C, *i.e.* EtAc–Cy (50 : 50); this yields a separation into 8 groups of less polar oestrogens, which are then characterised in different solvent systems, partly in the form of various derivatives such as Girard-

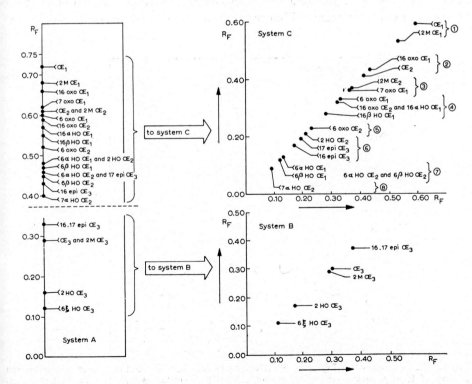

Fig. 4. System for the separation and characterisation of 24 steroid oestrogens on thin silica gel G layers; composition of systems and procedure *cf.* text. OE_1 = oestrone, OE_2 = oestradiol-17β, OE_3 = oestriol, M = methoxy, HO = hydroxy.

References p. 86

complexes, nitroso complexes or reduction products. As an example, let us consider the further separation of group 4 consisting of 6-oxo-oestrone, 16-oxo-oestradiol, 16α-hydroxy-oestrone, and 16β-hydroxy-oestrone. Fig. 5 shows the R_F-values of the reduction products obtained from (1): 16α-hydroxy-oestrone → oestriol, (2): 6-oxo-oestrone → 6α-hydroxy-17β-oestradiol, (3): 16-oxo-17β-oestradiol → 16-epi-oestriol, (5): 16β-hydroxy-oestrone → 16-epi-oestriol (again) and (4): a mixture thereof; on the right are the R_F-values of various reference compounds. In this manner each group is separated and characterised. So far this is a more or less "theoretical" example, since this separation scheme is obviously working with pure substances; whether it also works in a satisfactory way with urinary extracts, remains to be proven. In any event a thorough prepurification of the extracts would be necessary.

This is less important with regard to the estimation of the relatively high concentrations of urinary pregnanediol according to WALDI[9], which I should like to quote as a second example. Urinary pregnanediol is a useful indicator of endogenous progesterone production in the female and may serve as a control of the menstrual cycle or of early pregnancy, when the progesterone and pregnanediol level increases significantly.

After acid hydrolysis of the urine to split pregnanediol glucuronide, the hydrolysate is extracted with an equal volume of cyclohexane, a solvent which is of sufficient polarity to extract pregnanediol but not so polar as to extract many impurities; the

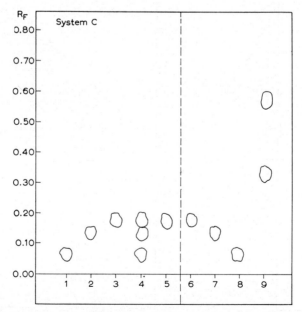

Fig. 5. TLC of group 4 (cf. Fig. 4) in ethylacetate–cyclohexane 50 : 50; (1): oestriol as reduction product of 16α-hydroxy-oestrone, (2): 6α-hydroxy-17β-oestradiol as reduction product of 6-oxo-oestrone, (3): 16-epi-oestriol as reduction product of 16-oxo-17β-oestradiol, (4): mixture of 1–3, (5): 16-epi-oestriol again as reduction product of 16β-hydroxy-oestrone, (6–9): reference compounds.

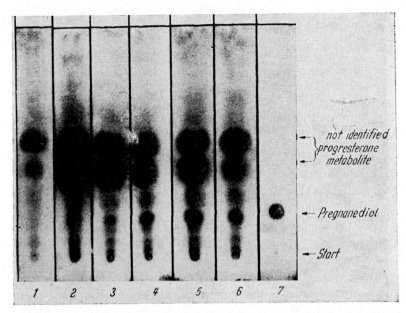

Fig. 6. Thin-layer chromatograms (1–6) of urinary extracts on the 4th, 8th, 13th, 16th, 21st and 24th day of the menstrual cycle.

extract is washed with sodium hydroxide and water and evaporated, and is then ready for TLC. Fig. 6 taken from STAHL's book, shows a series of chromatograms run with chloroform–acetone (9 : 1) in about 30 minutes and developed with phosphoric acid and phosphomolybdic acid; 1 to 6 represent extracts from samples of urine taken on approximately every 4th day during the cycle with increasing pregnanediol excretion until the 21st day (chromatogram 5). Large amounts of other metabolites and impurities are well separated, but low levels of pregnanediol as in chromatogram 1 and even 2–3 cannot be identified satisfactorily by this simple procedure.

5. CRITICAL REMARKS

Instead of presenting more examples which may be found in the literature or in the excellent books of STAHL, RANDERATH or TRUTER[10], I would prefer to discuss some of the relative merits of thin layer and paper chromatography with a view to facilitating the choice between these two techniques. Though I am a great friend of TLC — we are in fact using it with all types of natural and synthetic products, in particular with amino acids and peptides — there are quite a few problems for which I am by no means prepared to abandon paper chromatography. The question when to use paper or thin-layer chromatography cannot be answered generally without knowing the specific problem. However, Table 4 may provide some clue with regard to steroid chromatography — for example, particularly where the complex and relatively labile corticosteroids are concerned; in the field of corticosteroids, paper chromatography

TABLE 4
RELATIVE MERITS OF PAPER AND THIN-LAYER CHROMATOGRAPHY OF STEROIDS

Paper sheet	*Thin layer* (SiO_2, Al_2O_3)
favours the application of UV prints and specific reagents (*e.g.* alkaline fluorescence); sensitivity down to 0.1 µg/spot; favours difficult separations, in particular of more polar compounds, such as closely related corticosteroids.	favours the use of corrosive reagents for detecting unspecifically small amounts down to 0.01 µg/spot; direct UV print not possible; needs less time and material; somewhat more sensitive to impurities; gives cleaner eluates.

offers — in my opinion and according to my experience — more reliable methods both of separation and specific detection. I have mentioned earlier a few drawbacks of TLC when referring to methods of evaluation; furthermore, substances like aldosterone are much more labile on silica layers than on cellulose. If one has to perform hundreds of chromatograms a day it is obviously simpler just to use precut paper sheets than to prepare 50 plates or more. On the other hand, the sensitivity and speed of TLC is unsurpassed; this is an important advantage as regards many problems, including particularly the study of synthetic products, their transformation, and purity. Thus, in my view, these techniques supplement each other in an ideal fashion.

REFERENCES

1. E. HECKER, *Verteilungsverfahren im Laboratorium*, Supplement to Angewandte Chemie, Verlag Chemie, Weinheim, 1954.
2. E. STAHL, *Pharmazie*, 11 (1956) 633.
3. E. VON ARX AND R. NEHER, cited in ref. 4.
4. R. NEHER, *Steroid Chromatography*, Elsevier, Amsterdam, 1964.
5. Full references are to be found in ref. 4.
6. S. HEŘMANEK, V. SCHWARZ AND Z. ČEKAN, *Collection Czech. Chem. Commun.*, 26 (1961) 1669.
7. BR. BAEHLER, *Helv. Chim. Acta*, 45 (1962) 309.
8. B. P. LISBOA AND E. DICZFALUSY, *Acta Endocrinol.*, 40 (1962) 60.
9. D. WALDI, *Klin. Wochschr.*, 40 (1962) 827.
10. E. STAHL, *Dünnschicht-Chromatographie*, Springer, Berlin, 1962;
    K. RANDERATH, *Dünnschicht-Chromatographie*, Verlag Chemie, Weinheim, 1962;
    E. V. TRUTER, *Thin-Film Chromatography*, Cleaver-Hume Press Ltd., London, 1963.

THIN-LAYER CHROMATOGRAPHY OF LIPIDS

F. B. PADLEY

Department of Chemistry, The University, St. Andrews, Scotland (Great Britain)

1. INTRODUCTION

Thin-layer chromatography has become an essential technique to chemists working in the lipid field. Using this method, mixtures can be chromatographed in approximately half an hour to obtain separations much sharper than those given by paper or column chromatography. Only in the last five years has thin-layer chromatography become universally accepted however, although its history goes back much further.

The separation of mixtures of organic compounds on thin layers of powdered adsorbent was described by IZMAILOV AND SCHRAIBER[1] (1938) and WILLIAMS[2] (1947) who used a powdered adsorbent spread between two glass plates. The chromatogram was developed by introducing solute and solvent through a hole in the centre of the top plate.

A powdered adsorbent containing starch as a binding agent was used by MEINHARD AND HALL[3] (1949) to separate inorganic ions on coated microscope slides by radial development.

KIRCHNER, MILLER AND KELLER[4] (1951) described the first chromatostrips as we know them today, the solute being applied at one end of the strip which was then developed by the ascending technique. These investigators published numerous papers on the separation of terpenes and essential oils on chromatostrips, or on sheets for two-dimensional separations, and described an apparatus for preparing chromatoplates.

REITSEMA[5] (1954) extended the technique by making larger "chromatoplates" (5" × 7") and used them to separate essential oils. Chromatoplates have the advantage over strips that the uncertainty of identifying components is overcome by running known components alongside the unknown or as a mixture.

The wider applications of thin-layer chromatography were not visualised until 1956 however, when STAHL[6] described an apparatus for preparing chromatoplates and how they might be used to separate compounds other then terpenes.

Thin-layer chromatography is now used either alone or in conjunction with other chromatographic techniques to separate practically all the lipid classes.

Several papers[7-23] notably by STAHL[18] and MANGOLD[12] and books by RANDERATH[24], TRUTER[25] and STAHL[26] have reviewed thin-layer chromatography.

This present paper reviews the thin-layer chromatography of lipids excluding steroids.

References p. 111 (authors) and p. 115 (apparatus etc.)

2. PREPARATION OF CHROMATOPLATES

In order to obtain reliable separations and to aid the detection of separated components the thickness of the adsorbent layer should be as uniform as possible. Numerous commercial applicators are available and many less expensive methods of applying layers either to glass plates or strips have been described.

Kirchner[27–32] first described an apparatus for coating glass strips, and Reitsema (ref. 5, 33) prepared chromatoplates by applying a slurry of the powdered adsorbent in water with a spray gun. Stahl[6,34,35] described an applicator which is now commercially available (1), the principle of which is used in many later applicators.

The applicator contains a slurry of the adsorbent in water which leaves through a narrow slit made between the apparatus and the glass on which it rests. A uniform layer of adsorbent is spread by drawing the applicator smoothly across the plate. Five plates are prepared simultaneously by arranging the plate in line on a plastic board with a retaining edge. Before the slurry sets, the plates should be shaken gently to remove any small irregularities in the layer, dried in an oven and stored in a desiccator.

Layers of different thickness can be prepared with a modified version of Stahl's original applicator (1).

Another commercially available apparatus (2), based on Stahl's model but simpler in design, gives a layer approximately 275μ thick.

These applicators have the disadvantage that the glass plates used in each run must be of uniform thickness otherwise when the applicator is drawn across the plates it will stop or tilt at each junction. This problem has recently been solved by Shandon Scientific Co. Ltd. (3) who have patented a levelling platform which enables the alignment in the same plane of chromatoplates or chromatostrips. An apparatus (4) described by Mutter and Hoffstetter[36] also overcomes the problem of using plates of unequal thickness. In this case the applicator is fixed and the plates are pushed underneath it; the thickness of the layer may be altered. A modified form of this instrument with details for its construction is described by Wollish, Schmall and Hawrylyshyn[22].

Numerous applicators which are easily made in the laboratory have been described. Reitsema[5] and more recently Bekersky[37] used a spray gun to apply a slurry of the adsorbent in water to chromatoplates or chromatostrips. Bekersky determined the thickness[6] of the sprayed layers and showed that their thickness deviated within 40 microns compared with 20 microns for spread layers. Layers have been spread by drawing the adsorbent across a plate with a glass rod, the thickness of the layer controlled by placing rubber tips on the end of the glass rod[38] or by placing adhesive or metal strips[39–42] along two opposite sides of the glass plate. Applicators based on Stahl's model but simpler in design have been described by Barbier et al.[43], Machata[44], Vioque[20], Ritter[45] and Mottier[46]. Ridged glass plates were used by Gamp[47] in order to obtain layers of reproducible thickness and to prevent the side flow of solvent.

Peifer[48] described a method of spreading layers on plates of varying sizes by dipping the plates into a suspension of the adsorbent in a volatile solvent (chloroform or chloroform/methanol). The plates were dried for 1–3 min over a hot plate before use. HOFMANN[49] and WASIKY[50] used a similar technique for chromatography on microslides.

A simple method of coating the inside of test tubes and their use in the chromatography of lipids has been published by LIE[51]. ROSI AND HAMILTON[52] coated both sides of their chromatoplates to obtain more efficient use of the plates.

The applicators described can generally be used to spread the adsorbent either as a slurry or as a dry powder.

3. TYPES OF LAYER

Chromatography on thin layers can be divided into three classes, adsorption, partition and ion-exchange chromatography. Important advances have been made in reverse-phase partition chromatography and in the chromatography on adsorbents modified by impregnation with reagents that form complexes with functional groups.

Silica gel

Silica gel, normally mixed with calcium sulphate to act as a binding agent, is widely used as an adsorbent on which lipids are separated into classes. WREN[53] has reviewed the chromatography of lipids on silica gel and these separations are generally applicable to thin-layer chromatography. KLEIN[54] discussed chromatography on silica gel in terms of surface energy and structure. The activation and standardisation procedures necessary to reproduce silica gels with all the Brockmann activity grades have been described by HERNANDEZ, HERNANDEZ JR. AND AXELROD[55]. These procedures are however normally unnecessary in thin-layer chromatography.

Silica gel (200 mesh) with or without a binding agent (15 % calcium sulphate or 5 % starch) and prepared specifically for thin-layer chromatography is commercially available (1–9). A similar adsorbent may be prepared[12] by thoroughly mixing silica gel (200 mesh Mallinckrodt) (10) with calcium sulphate (10–15 %, 200 mesh). Powdered adsorbents which do not contain a binding agent but adhere to the glass plates are also available (9).

The adsorbent (30 g) is normally spread as a slurry in water (60 ml) onto glass plates, the chromatoplates are then dried at 110° C for one hour and stored in a desiccator ready for use.

The adsorption characteristics of silica gel may be modified by impregnating it with reagents. The separation of acidic components is improved by chromatography on an acidic adsorbent prepared by slurrying the adsorbent in an acidic solution. STAHL[56] used a solution of oxalic acid (0.5 N) and PEIFER AND MUESING[57] used a chloroform/methanol (3 : 2) solution of sulphuric acid (2.5 %). Phospholipids and strongly acidic fatty acid derivatives were separated on layers prepared with a slurry of silica gel G (25 g) in aqueous ammonium sulphate (60 ml, 4 % soln.) by MANGOLD AND KAMMERECK[58]. STAHL[56] slurried silica gel G in aqueous potassium hydroxide (0.5 N) and

References p. 111 (authors) and p. 115 (apparatus etc.)

carotenals were separated by WINTERSTEIN et al.[59] who prepared chromatoplates with a slurry of silica gel G (5 g) and calcium hydroxide (20 g) in water (50 ml). SKIPSKI, PETERSON AND BARCLAY[60] prepared a basic adsorbent by slurrying silica gel G in sodium acetate (0.01 M) or sodium carbonate (0.01 M) and separated phosphatidyl serine from other phosphatides.

Some closely related compounds and isomers have recently been separated by chromatography on silica gel G impregnated with complexing reagents. HALMEKOSKI[61] treated silica gel with the chelate forming anions, molybdate, tungstate or borate, to improve the separation of phenolic carboxylic acids. MORRIS[62] sprayed silica gel G chromatoplates with saturated methanolic solutions of either silver nitrate or boric acid. Using the dried plates he separated methyl esters of unsaturated fatty acids or hydroxy compounds respectively. BARRETT, DALLAS AND PADLEY[63] prepared an adsorbent of silica impregnated with silver nitrate by slurrying silica gel G (30 g) in a solution (60 ml) of silver nitrate in water (12.5%) and separated glycerides according to their degree of unsaturation. DE VRIES[64] who originally described the separation of methyl esters and triglycerides by column chromatography on a silica/silver nitrate adsorbent has also separated methyl esters and triglycerides on thin layers of this adsorbent[65].

More recently, MORRIS[66] impregnated silica-gel layers with boric acid, sodium borate or sodium arsenite and on these separated isomeric polyhydroxy fatty acid esters.

Alumina

Vitamins, carotenoids and lipids have been chromatographed on alumina. Its use as an adsorbent is limited however because certain organic compounds are modified during chromatography. E. LEDERER AND M. LEDERER[68] discussed the modification of organic compounds on alumina or silica gel. The modification of terpenoid hydrocarbons[67], the hydrolysis of ester linkage[69] and the isomerisation of double bonds[69-72] on alumina has also been investigated.

Several brands of alumina for thin-layer chromatography are available (3, 5, 6, 8, 9). The adsorbent normally contains calcium sulphate (5%) as a binding agent but numerous workers use the adsorbent with no bonding agent[38,39]. A mixture of Alcoa activated alumina (11) (200 mesh) with calcium sulphate (200 mesh, 5%) is satisfactory for thin-layer chromatography[183].

The chromatoplates are normally prepared by spreading a slurry of the adsorbent in an appropriate amount of water (varies depending on the make of alumina) and drying the plates in an oven to activate the alumina to the required level. The activity depends on the length of time the plates are dried at a given temperature. Brockmann grade II activity[73] is obtained by heating the plate at 200–220° C for four hours and grades III or IV by drying at 150–160° C for four hours[12]. The plates should be stored over alumina of the appropriate activity in a desiccator. STAHL[74] has described a method for determining the activity of alumina according to the Brockmann scale[73] by using a series of dyes. The necessary equipment is commercially available (1).

Cellulose powder

Alkaloids[75] and more recently amino acids[76] have been separated on cellulose powder (7).

TEICHERT, MUTSCHLER AND RÖCHELMEYER[75] modified cellulose by impregnating the layer with a solution of formamide in acetone (20%). WAGNER[77] separated phospholipids on this modified adsorbent. Highly acetylated cellulose powder spread on glass plates was used by WIELAND, LÜBEN AND DETERMANN[78]. Partially acetylated cellulose was recently used by BADGER et al.[79]. Gels of fractionated dextrans (12) were used by DETERMANN[80] and WIELAND AND DETERMANN[81] to separate compounds of high molecular weight e.g. peptides and proteins.

Impregnation of layers for reverse-phase chromatography

Individual members of a lipid class are separated from one another by reverse-phase partition chromatography e.g. MANGOLD AND MALINS[82] separated fatty acid methyl esters using silicone oil as a stationary phase on silica gel G.

The layer for reverse-phase chromatography is normally prepared by slowly immersing a dry chromatoplate of silica gel G[82], kieselguhr[83] or gypsum[84] in a solution of the stationary phase in light petroleum. The temperature of the chromatoplate and the solution should be the same because the adsorbent layer will disintegrate if a warm plate is dipped into the cooler solution. Numerous stationary phases have been used notably by KAUFMANN et al.[82-90] who impregnated chromatoplates with light petroleum solutions of undecane (15%) tetradecane (5%) (Fig. 1), silicone oil, 50 cp. (7%), paraffin oil (5%) and high boiling petroleum, 240–250° C (5%). Tetradecane,

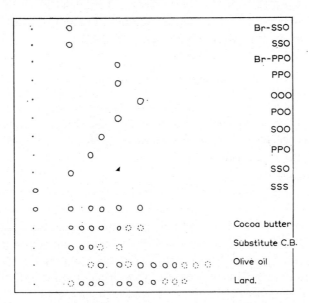

Fig. 1. Separation of brominated triglycerides on kieselguhr impregnated with tetradecane. Developed with propionic acid/acetonitrile, 6 : 4, saturated (80%) with tetradecane + 0.5% bromine. (Kaufmann, Makus and Khoe, 1962.)

undecane and high boiling petroleum have the advantage that these phases can be evaporated after chromatography thus enabling the detection by normal methods of the separated components[83].

MANGOLD[12,91] used polyethylene powder (Hostalen S[13]) to separate fatty acids or their methyl esters.

Dicarboxylic acids were separated on layers of silica gel/polyethylene glycol by KNAPPE[92], STAHL[93] used Perlon (polyacrylonitrile) as a stationary phase and BIRKHOFER AND KAISER[94] separated anthocyanius cinnamic acids and simple sugars. Antioxidants were separated on layers of polyamide powder by DAVIDEK AND PROCHAZKA[95].

Ion-exchange chromatography

Modified celluloses (7) have been used for thin-layer chromatography. RANDERATH[96] separated nucleotides on diethylaminoethyl (DEAE) cellulose and WIELAND AND DETERMANN separated LDH isozymes on DEAE sephadex[81]. WEIMANN AND RANDERATH[98] recently used cellulose powder impregnated with nondialysable polyethyleneimine and claim that it has certain advantages over DEAE cellulose.

Other adsorbents

Kieselguhr G (1, 3, 5, 8) has been used to fractionate lipid extracts[96] and sugars[97]. The early work of KIRCHNER[27] describes the preparation of thin layers of magnesium oxide, magnesium carbonate, calcium hydroxide and calcium carbonate. BRODASKY[100] has recently separated neomycin sulphates on carbon. Hydroxyl apatite was used by HOFMANN[101] to separate partial glycerides. PEIFER[48] used Florisil[102] as a coating for microchromatoplates. HIRSCH[103] recently separated nonpolar lipids by column chromatography on factice (14) (a polymerised vegetable oil) and this will probably be a suitable adsorbent for thin-layer chromatography.

4. DEVELOPMENT OF CHROMATOPLATES

A solution of the mixture (0.1–1 %) in light petroleum or chloroform is applied either as a spot (500 g/spot) or narrow band to the plate with either, a capillary drawn out from glass tubing, micro-pipette or micro-syringe. If possible the layer should not be damaged during the application and the spots should be as small as possible to obtain the sharpest separations.

The plate may be developed vertically or horizontally with an appropriate solvent. The choice of solvent will depend on the nature of the mixture to be separated. BROCKMANN AND VOLPERS[104] showed that in column chromatography the polarity of a compound depends on the number and type of functional groups it contains and this rule was shown by STAHL[105] to apply in thin-layer chromatography. Saturated hydrocarbons are weakly adsorbed and the adsorption of unsaturated hydrocarbons increases with the number of double bonds and with conjugation. The introduction of other functional groups causes adsorption to increase in the following order $-CH_3 < $ O-alkyl $ < C=O < -NH_2 < OH < -COOH$. Eluotropic series have been devised[70-72,106,107] notably by TRAPPE, in which a series of solvents are arranged in order of polarity and eluting power *i.e.* light petroleum, cyclohexane, carbontetrachloride, trichloroethylene, toluene, benzene, dichloroethane, chloroform, diethylether, ethylacetate, acetone, isopropanol, ethanol, methanol. Mixed solvent systems are generally used to give optimum resolution. BRENNER AND PATAKI discussed some theoretical aspects of thin-layer chromatography[108].

The choice of solvent is thus simplified with knowledge of the polarity of the mixture to be separated used in conjunction with the eluotropic series. Numerous solvent systems have to be screened however and this is best done using chromatostrips or microslides[49]. IZMAILOV[1] used a rapid circular development method, placing drops of the solvent onto a spot of the mixture to be separated on the plate. The components were separated in concentric rings and the most suitable solvent was that giving the clearest separation.

The plates are normally developed by the ascending technique in an enclosed tank containing solvent. When layers of loose powder are used the plates are developed either horizontally or inclined to a slight angle[109-116].

Lipids have been separated by gradient elution which enables the separation on the same plate of compounds with very different polarities[117-122]. The chromatoplates are usually developed at room temperature, however low temperature chromatography was employed by MALINS AND MANGOLD[82] to separate critical pairs *e.g.* palmitic and oleic acid, by reverse-phase chromatography.

The Matthias wedge[121] strip technique was used by STAHL[35] and COPIUS PEEREBOOM[16] to improve the resolution of compounds from one another. The effect of the width and thickness of the chromatostrip on R_F values was discussed by FURUKAWA[122].

Chromatoplates have also been developed several times either with the same solvent system or different solvent systems[56]. Multidevelopment with different solvent systems has been used to remove the less polar lipids from a mixture prior to separating the polar components. WEICKER[123] separated total serum lipids using three successive solvent systems. Multidevelopment using the same solvent system was first used in paper chromatography to obtain better resolutions of compounds with similar R_F values[124]. KAUFMANN[89] applied this technique to separate critical pairs of unsaturated triglycerides and THOMA[125] described the application and theory of unidimensional multiple chromatography. Accelerated chromatography by centrifugation is well known in paper chromatography and no doubt will be of great value for multi-developing chromatoplates. CHAKRABARTY[117] centrifugally accelerated the development of chromatoplates and more recently HERNDON *et al.*[127] described an apparatus for centrifugally accelerating horizontal chromatography on chromatostrips or paper.

Horizontal continuous running thin-layer chromatography was described by BRENNER AND NIEDERWIESER[110] and STANLEY, IKEDA AND COOK[111]. More recently MISTRYUKOV[115] devised a method of continuous elution by descending development on layers of non-bound powdered alumina. Better resolution between components with similar R_F values was obtained and solvents which gave low R_F values but had a high elution power could be used.

References p. 111 (authors) and p. 115 (apparatus etc.)

Complex mixtures of lipids may be separated into classes and then the components within each class separated by two-dimensional chromatography. A single spot of the complex mixture is placed at one corner of the plate, and after development the plate is turned through 90° and redeveloped either in the same solvent or a different one. If a mixture is chromatographed in both directions with the same solvent system, better resolution will be obtained then for a single development; the components, providing they are not being modified during chromatography, will lie on a diagonal line. KAUFMANN AND MAKUS[85] separated complex mixtures by adsorption chromatography in one direction and reverse-phase chromatography in the second. The components on the plate have also been modified prior to the second development. PASTUSKA AND TRINKS[128] separated naphthols, amines and dyes and RAMSEY[129] separated esterases by thin-layer electrophoresis.

R_F values determined by the ratio of the distance moved by a component to the distance travelled by the solvent front are generally not as reproducible as in other chromatographic methods. The ratio of the distance moved by the component compared to the distance travelled by a reference compound is more useful. For identification purposes however, known compounds should be chromatographed alongside the unknown. STAHL[56] noticed that if mixed solvent systems were used and in particular if they were of very different volatility there was a tendency for the R_F values to be higher at the outside than at the centre of the plate. This "border effect" was eliminated by lining the chromatographic tank with filter paper saturated in the solvent. Development was also quicker and a more polar solvent was required but the separations were not as sharp as those obtained using an unlined tank.

SEHER[131] developed chromatoplates with chloroform prior to chromatography, and obtained uniform R_F values.

The reproducibility of R_F values in thin-layer chromatography has been investigated by BRENNER et al.[132] who concluded that the purity of the solvents and the degree of vapour saturation in the chromatographic tank had a very great effect on R_F values. The effects of temperature, thickness of the layer, quality of the adsorbent, elution time, point of application of the mixture and type of elution were also discussed. Taking suitable precautions they obtained R_F values which varied within the same order of magnitude as those obtained by paper chromatography.

5. MODIFICATION OF MIXTURES TO IMPROVE SEPARATIONS

The conversion into derivatives of compounds in a mixture sometimes simplifies their separation by chromatography. The components may be modified before or after applying the mixture to the chromatoplate and this latter technique has been widely applied in two-dimensional chromatography.

The unsaturated components in a mixture may be oxidised during chromatography by placing peracetic acid in the developing solvent[82,133,134]. In this way MALINS AND MANGOLD[82] separated saturated from unsaturated nitriles and critical partners such as palmitic and oleic acid.

The preparation of mercuric acetate addition products of unsaturated compounds was described by JANTZEN AND ANDREAS[135], and MANGOLD AND KAMMERECK[136] separated the mercuric acetate adducts of unsaturated acids by thin-layer chromatography. Radio-active labelled lipids were prepared by MANGOLD[134,137] by reacting their functional groups with either ^{14}C-labelled diazomethane or ^{14}C- or ^{3}H-labelled acetic anhydride; ^{3}H-labelled diazomethane cannot be used for methylation[138,139]. The separated lipids can be estimated by radiometry.

In preparative work it is usually advantageous to convert alcohols and acids to acetates or methylesters to make recovery of the separated components easier[137].

MILLER AND KIRCHNER[140] described the oxidation, reduction, dehydration, hydrolysis and the formation of derivatives on the plate. Recently KAUFMANN, MAKUS AND KHOE[90] hydrogenated or brominated unsaturated acids and glycerides on the chromatoplate. Hydrogenation was effected by spraying the strip of adsorbent containing the applied mixture with a colloidal solution of palladium (2%). The plate was dried in a desiccator. For complete hydrogenation the plate was impregnated with a solution of undecane (10%) in light petroleum prior to standing in an atmosphere of hydrogen for one hour. Unsaturated components were brominated by placing bromine (0.5%) in the developing solvent. STAHL[141] described a Separation, Reaction, Separation (S.R.S.) technique whereby chemical reactions were studied by two-dimensional chromatography.

6. DETECTING COMPONENTS

Detecting components on chromatoplates is simplified to a certain extent because the layer is not effected by corrosive reagents. Practically all the reagents used to detect components on paper chromatograms may be used for detection in thin-layer chromatography[142,143]. The layer is easily damaged by excessive spraying and immersing in aqueous reagents however.

This problem has recently been overcome by KAUFMANN[87] who placed the plate in an atmosphere of dichloromethylsilane, silanising the silica layer which was then strong enough to be immersed in reagents as in paper chromatography. Thus, fatty acids were detected as copper salts with rubeanic acid; gypsum layers[84] (8) are also stable to immersion in aqueous solutions.

Besides the more specific reagents for detecting compounds on the plate three common reagents are:

(1) Iodine vapour. The chromatoplate is placed in a tank containing a few crystals of iodine to detect unsaturated and usually saturated compounds. TIMS[144] investigated the limits of sensitivity of iodine vapour to a wide range of lipids.

(2) 2′,7′-Dichloro- or dibromofluorescein (0.2% in 95% ethanol). The plate is sprayed with this reagent and the components are seen as yellow fluorescent spots against a dark background when viewed in ultraviolet light.

(3) Corrosive reagents[4,174]. Organic materials are charred by spraying the plate with aqueous sulphuric acid (50%) and heating at approximately 200° C. This method is limited in its use because the components are destroyed.

References p. 111 (authors) and p. 115 (apparatus etc.)

Spray reagents for general and specific detection have been described by MANGOLD[12], MANGOLD AND MORRIS[194] and WOLLISH et al.[22]. Specific reagents for phosphatides and their hydrolytic products were described by SKIDMORE AND ENTENMAN[202].

7. RECORDING CHROMATOGRAMS

A simple but tedious method of recording a thin-layer chromatogram is to trace it onto tracing paper but only a general outline of the separation is obtained.

MEINHARD AND HALL[3] removed the silica layer from the chromatoplate with scotch tape. BARROLLIER[145] and LICHTENBERGER[146] described methods of converting the adsorbent layer into a stable plastic film. The plates were either dipped into label glaze solution or sprayed with a dispersion in water of vinylidene chloride (15 %) or vinylpropionate (15 %), which polymerises on the plate. The stable layer was then stripped from the plate after immersion in water. A similar effect is achieved by spraying the plate with Neatan (5, 8).

SEHER[131] exposed sheet film or photographic paper in an enlarger and EISENBERG[147] has recently recorded chromatograms using diazo paper (15). The paper can be handled freely in normal laboratory light and was exposed by placing the paper onto the adsorbent layer and holding over a table lamp for ten minutes. The lampshade acted as a suitable support. The print was developed in a tank containing an open beaker of ammonia. Unexposed areas appeared as bright blue spots and the contrast was better than on the original plate.

REITSEMA[5] photographed chromatoplates and BÜRKI AND BOLLINGER[148] and ABELSON[149] photographed chromatoplates in normal or ultra-violet light. The Polaroid camera has also been used[150,151].

Chromatoplates have been photocopied by numerous workers[152-157] and the photocopies used in quantitative analysis[131,152-154,158].

Chromatoplates of radio-active labelled lipids have been reproduced on X-ray prints by MANGOLD[12,134].

8. QUANTITATIVE THIN-LAYER CHROMATOGRAPHY

Complex mixtures can be quickly separated by thin-layer chromatography and in many cases the separations cannot be reproduced by any other method. Quantitative methods of estimating the separated components have therefore been devised.

Semi-quantitative estimations were obtained[159] by eluting the components from the plate and weighing. To overcome the inaccuracy incurred in weighing small quantities colorimetric methods of estimation have been developed. GÄNSHIRT et al.[162] estimated bile acids colorimetrically after elution with aqueous sulphuric acid and heating. HABERMANN, BANDTLOW AND KRUSCHE[163] and WAGNER[77] estimated phospholipids as phosphate after elution and mineralisation. Ubiquinones were quantitatively estimated[164] and also flavenoids[160] by elution and colorimetry. VIOQUE AND HOLMAN[165] separated esters of different types; the spots were removed, extracted and converted to iron-hydroxamic acid complexes which were estimated colorimetrically. The recovery of esters from the plate was better than 95% and estimation was ± 2.5%.

Radioactive-labelled lipids were separated, eluted and estimated by radiometry by MANGOLD[58] and BROWN AND JOHNSTON[166]. SNYDER AND STEPHENS[126] did not elute the spots and suspended the adsorbent in the scintillation solution.

When mixtures are being analysed by the above methods one should ensure that the more polar components have been completely eluted.

Methods which directly determine components on the plate have been described. HEYNS[167] used a mass spectrometer to estimate substances provided that > 1% was present in the mixture and the spots were sharp.

FISHER et al.[168] showed that the size of the spot on a paper chromatogram was directly proportional to the amount of material present. SEHER[153,154] suggested that a similar relationship existed for thin-layer chromatograms and made use of this principle to estimate tocopherols separated by T.L.C. BRENNER AND NIEDERWIESER[169] and STAHL[19] investigated the relationship between spot area and concentration and found that at low concentration a linear relationship between weight and area existed. PURDY AND TRUTER[170] recently demonstrated that in thin-layer chromatography the square root of the area of a spot is a linear function of the logarithm of the weight of the material in the spot. Two methods which depended on this principle were described for quantitatively estimating separated components. The standard deviation was 2.6% when determining n-hexadecanoic acid and n-hexadecanol. Even greater accuracy was possible in reverse-phase chromatography because the spots were larger. Photographs of photostats of the chromatoplates have been used for quantitative analysis[131,152-154].

The measurement by photodensitometry (16) of the charred components on the chromatoplate has been widely applied by PRIVETT, BLANK AND LUNDBERG[171-173]. The method is simple once the basic manipulative skills have been acquired and is particularly useful for analysing multicomponent systems. The separated components are destroyed during the estimation however but in the majority of cases the γ-quantities of material required for the analysis are expendable.

PRIVETT, BLANK AND LUNDBERG found that when the components were charred by spraying with sulphuric acid (50% aqueous) and heating at 200° C, the density of the spots was directly proportional to the weight of the component but different classes of lipids gave spots of different intensity. (Standard correction factors were determined and did not vary appreciably from one plate to the other.) Recently the same workers[174] described charring conditions in which unsaturation had little effect on the intensity of the spot. The plate was sprayed lightly with a saturated solution of potassium dichromate in sulphuric acid (80% aqueous by wt.) and heated at 180° C for 25 min.

Simple mixtures of triglycerides and partial glycerides and also glyceryl residues from the ozonolysis of mixtures of saturated and unsaturated mono-, di- and triglycerides were estimated by photodensitometry after charring[171]. The four types of

References p. 111 (authors) and p. 115 (apparatus etc.).

monoglycerides, six of the seven types of diglycerides and four of the six possible triglycerides were analysed.

A photodensitometric method was used by BARRETT, DALLAS AND PADLEY[175,176] to determine the composition of unsaturated triglyceride mixtures separated on a silica/silver nitrate adsorbent. Four chromatograms of the unknown mixture were separated alongside a similar number of standard mixtures on the same plate. The components were charred at 25° C for 5 min after spraying with phosphoric acid (50% aqueous). The density of the components was measured (17) and correction factors, determined from the standard mixtures, were applied to the unknown mixture. Separate correction factors were determined for each class of compound and correction factors varied from plate to plate.

Some instruments used for measuring the density of spots on paper strips are not readily adaptable for scanning chromatoplates. CSALLANY[177] removed the adsorbent layer after spraying with Neatan (8) solution and then sprayed the reverse side. The layer was cut into strips and scanned as in paper chromatography. BARRETT et al.[175,176] cut the glass chromatoplate after development into 1″ strips which were scanned by the "Chromoscan" photodensitometer (17).

9. PREPARATIVE THIN-LAYER CHROMATOGRAPHY

Up to 50 mg of a lipid mixture can be separated on a single chromatoplate (250–275 μ thick) and larger amounts have been separated by chromatography on thicker layers[41,45,178] of adsorbent and by using a number of chromatoplates.

The mixture to be separated should preferably be applied as a row of spots a small distance apart to help the development. A simple apparatus for applying simultaneously the required number of spots to chromatoplates was devised by MORGAN[179] and DUNCAN[41] applied with a syringe a solution of a mixture as a narrow band.

The separated components are usually detected by spraying one edge of the chromatogram with a non-destructive reagent such as 2,7-dichlorofluorescein. VIOQUE AND HOLMAN[165] found that when fatty components were detected with iodine the fatty-acid analysis by gas-liquid chromatography of the extracted material indicated a decrease in the unsaturated fatty-acid content.

The adsorbent containing each component is removed from the plate and extracted with solvent. RITTER[45] described a device working on the vacuum cleaner principle to collect adsorbent from the plate. Solvents which carry the compound to the solvent front in thin-layer chromatography, are suitable to extract the material from the adsorbent[82]. Solvents which dissolve calcium sulphate should be avoided unless the adsorbent contains no binding agent. Chromatography on layers of powdered adsorbent without a binding agent therefore has the advantage that there is little danger of contaminating the product with inorganic material; the adsorbent is more easily removed from the plate also. Highly polar lipids are not isolated in good yields from the adsorbent and wherever possible the less polar derivatives e.g. acetates or methyl esters[137] should be chromatographed.

ENER[180] and BACHLER[181] isolated volatile components by holding a cool glass plate above the heated chromatoplate. When horizontal development is used[109-116] the components can be isolated by allowing the solvent to run off the edge of the plate[12].

MANGOLD, KAMMERECK AND MALINS[137] analysed commercial preparations of radio-active labelled lipids and isolated milligram amounts. MANGOLD AND KAMMERECK[136] separated the mercuric acetate adducts of unsaturated fatty acids and analysed the regenerated acids by gas-liquid chromatography. This method enabled the resolution of overlapping peaks on the gas-liquid chromatogram. JENSEN et al.[182] used the same technique to analyse the fatty acids of butter. Preparative thin-layer chromatography has been used to isolate fatty acid methyl esters from other lipids[183]. PRIVETT AND BLANK[184] recently isolated by this method products formed during the initial stages of autoxidation of unsaturated fatty acids methyl esters.

Methods of transferring to column chromatography the separations obtained by thin-layer chromatography have been described. DAHN AND FUCHS[185] prepared a column of adsorbent (1000 × wt. of mixture) in a cellophane tube and developed the chromatogram horizontally as in thin-layer chromatography. DUNCAN[41] separated mixtures on a column prepared from adsorbent 500 to 1000 times the weight of the mixture being separated. The column was eluted slowly with solvent and a large number of fractions were taken. The criterion for separation on a column was given by:

$$r = \frac{a}{b + 0.1a}$$

where $a = R_F$ of fastest moving component and $b = R_F$ of slowest moving component; if $r =$ unity, resolution will occur.

10. THE SEPARATION OF LIPIDS

The separation of lipids into classes

Lipid mixtures are separated into classes by chromatography on layers of silica gel G developed usually with mixtures of light petroleum and diethyl ether with added acetic acid (0.5 %). Separation is according to the type and number of functional groups and is marginally affected by chain length and unsaturation (Fig. 2).

MANGOLD AND MALINS[82,186] fractionated complex lipid mixtures *i.e.* animal and vegetable lipids, fish oils and the unsaponifiable matter of lanolin, on silica gel G and on layers impregnated with silicone. KAUFMANN AND MAKUS[85] separated standard mixtures of lipids by adsorption and reverse-phase chromatography.

WEIKER[123] examined serum lipids by thin-layer chromatography and developed a chromatoplate with three solvent systems successively to separate cholesterol and its esters, lecithin, carotene, several fatty acids and unidentified compounds. JATZKE-WITZ[187] used eleven solvent systems when separating brain lipids and VOGEL[188] developed a number of individual chromatograms, each with different solvents, on the same plate when examining serum lipids. MANGOLD AND TUNA[189] analysed lipids extracted from human tissues and WILLIAMS[159] examined human fecaliths. HUHN-

References p. 111 (authors) and p. 115 (apparatus etc.)

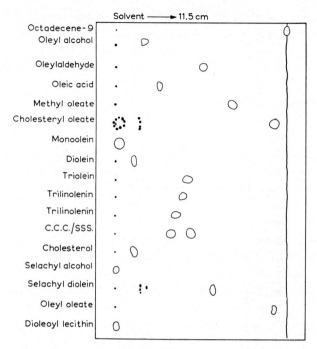

Fig. 2. Thin-layer chromatography of lipids on silica gel G. Solvent: light petroleum/ether/acetic acid, 90/10/1. (Malins and Mangold, 1960.)

STOCK AND WEICKER[161] analysed human serum lipids by two-dimensional chromatography using a two-phase solvent system (propylalcohol/ammonia, 2 : 1) in one direction and carbontetrachloride in the other. The lipids were semi-quantitatively estimated after separation.

VOCIKOVA[39] chromatographed serum lipid fractions on layers of unbound alumina. The plate was developed first with a light petroleum/ether (95 : 5) mixture separating cholesterol esters from triglycerides and leaving the more polar components on the base line. The adsorbent containing the separated components was replaced with fresh adsorbent and the plate redeveloped with a mixture of light petroleum/ether/acetic acid (94.5 : 5 : 0.5) separating free fatty acids, phospholipids and cholesterol.

LIE AND MYE[51] separated lipids on layers of adsorbent coated on the insides of test-tubes.

Fatty Acids

(i) Reverse-phase chromatography. MALINS AND MANGOLD[82] resolved saturated and unsaturated long chain fatty acid methyl esters according to chain length. One double bond in the acid had the effect of decreasing the apparent chain length of the acid by two methylene groups. Thus methyl palmitate and methyl oleate had the same R_F values at room temperature. KAUFMANN AND MAKUS[85] separated saturated and unsaturated acids on layers of silica gel G impregnated with undecane. Undecane had

the advantage that it could be evaporated after development making detection of the separated components easier.

Critical pairs of fatty acid methyl esters were resolved by MALINS AND MANGOLD[82] by low-temperature chromatography on siliconised chromatoplates or by developing the plates with an oxidising solvent[82,133,134]. KAUFMANN, MAKUS AND KHOE separated critical pairs of fatty acids by reverse-phase chromatography on layers of kieselguhr G[90] or gypsum[84] impregnated with undecane after hydrogenating or brominating the unsaturated acids on the plates. The homologous series of dicarboxylic acids, from oxalic to sebacic acid, was separated by KNAPPE AND PETERI[92] in silica gel–poly-ethylene glycol (M 1000), developed with di-isopropyl ether/formic acid/water (90 : 7 : 3).

(ii) Separation according to degree of unsaturation. Reverse-phase chromatography separates acids according to their degree of unsaturation, but the separations are complicated because they are dependent on chain length also. The separation of fatty acid methyl esters according to degree of unsaturation only, has been achieved in two ways.

MANGOLD AND KAMMERECK[136] converted the unsaturated acids into mercuric acetate adducts following the procedure of JANTZEN AND ANDREAS[135]. The adducts were separated from the unreactive saturated components and from each other depending on the numbers of double bonds in the original acid (Fig. 3). Saturated esters were separated from mercuric acetate adducts by developing the plate (silica gel G) with a light petroleum/ether (4 : 1) mixture. The mercuric acetate adducts were separated from one another by developing the plate in the same direction with a mixture of propanol/glacial acetic acid (100 : 1). The adducts were recovered from

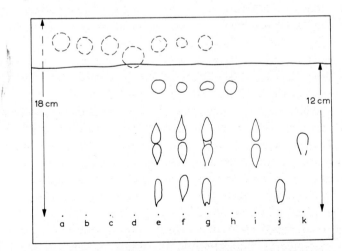

Fig. 3. T.L.C. of mercuric acetate adducts on silica gel. (Mangold and Kammereck, 1961.) 1st solvent: pet. ether/ether (18 cm) (80/20), 2nd solvent: *n*-propanol/acetic acid (12 cm) (100/1). Methyl esters: a. stearate, b. oleate, c. linoleate, d. linolenate, e–j. mercuric acetate adducts of: e. C_{16} fraction from Chlorella, f. C_{18} fraction from Chlorella, g. total methyl esters from Chlorella Pyrenoidosa, h. oleate, i. linoleate, j. linolenate, k. mercuric acetate. Indicators: ------ iodine, ———— sym-diphenylcarbazone.

References p. 111 (authors) and p. 115 (apparatus etc.)

the plate and the unsaturated acids easily regenerated unchanged from analysis by gas–liquid chromatography. JENSEN AND SAMPUGNA[182] analysed the fatty acids of butter by this method. This preliminary fractionation made it possible to separate acids which previously overlapped on the gas–liquid chromatogram.

Because mercuric acetate reacts twenty times faster with *cis* than *trans* ethylenic bonds it is possible to separate such isomers. A mixture of *cis*- and *trans*-isomers is reacted with mercuric acetate in an amount equivalent only to the *cis*-unsaturated components. The mercuric acetate adduct of the *cis*-olefin is readily separable by thin-layer chromatography from the unreacted *trans*-isomers[136].

Recently MORRIS[62] separated fatty acid methyl esters according to degree of unsaturation and certain *cis–trans*-isomers by chromatography on silica gel G impregnated with silver nitrate. DE VRIES[64] who first fractionated unsaturated fatty acid methyl esters on a column of silica impregnated with silver nitrate has also obtained similar separations by thin-layer chromatography[65] on a similar adsorbent. Thus the following mixtures of methyl esters were separable: stearate, elaidate, oleate, brassicate, erucate; positional isomers of octadecanoate; positional isomers of octadecadienoate; geometric isomers of 9,12-linoleate and geometric isomers of 9,12,15-linolenate. Solvent mixtures of diethyl ether/hexane and benzene/light petroleum are suitable for developing the chromatograms. This technique should prove extremely useful in studying the course of hydrogenation.

(iii) Separation of fatty acids containing other functional groups. Epoxy and hydroxy acids are separable by chromatography on silica gel G by developing the chromatogram with light petroleum/diethyl ether mixtures alone or containing acetic acid (1%).

MORRIS, HOLMAN AND FONTELL[152] isolated hydroxy and epoxy acids and demonstrated the interference of unsaturated hydroxy compounds in estimating epoxides[190]. The same workers[191–193] combined the techniques of paper chromatography, thin-layer chromatography and gas-liquid chromatography to analyse mixtures of epoxy and hydroxy acids and were able to detect 0.1% epoxy acid in a 1 mg-sample of mixed acids or esters. Epoxy acids (3.6%), monohydroxy acids (6%) and triterpenoid acids were detected in Orujo oil (sulphur olive oil)[192].

KAUFMANN AND MAKUS[85] separated epoxy, hydroxy, episulphido, and normal acids by adsorption and reverse-phase chromatography on silica gel G or silica gel G impregnated with undecane respectively. The plates were developed either with isopropyl ether/acetic acid (1.5%) or acetic acid/water (80 : 20). Keto acids, lactones and hydroxy acids were separated by reverse-phase chromatography[86] on kieselguhr impregnated with petroleum (240–250° C) and developed with glacial acetic acid/water (80 : 20) saturated with petroleum.

MANGOLD AND MORRIS[194] partially separated the stereoisomers of unsaturated hydroxy acids *e.g.* 9-hydroxy-*trans*-10-*trans*-12-octadecadienoic and 9-hydroxy-*trans*-10-*cis*-12-octadecadienoic acids by adsorption chromatography on silica gel G. The positional isomers of some epoxy, hydroxy and chloro-hydroxy unsaturated fatty acids were also separated by chromatography on silica gel G; a mixture of light petroleum/diethyl ether/acetic acid (70/30/2) was used to develop the chromatogram.

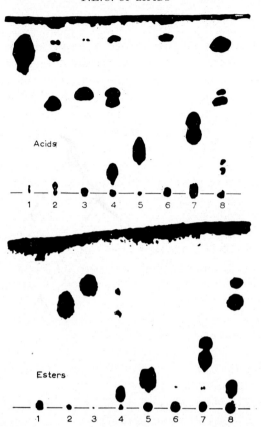

Fig. 4. T.L.C. of oxygenated fatty acids and esters on silica gel. (Mangold and Morris, 1962). Acids: pet. ether/ether/acetic acid (90/10/1); Esters: pet. ether/diethyl ether (90/10); 1. palmitoleic + oleic, 2. cis-9,10-epoxystearic, 3. cis-12,13-epoxyoleic, 4. cis-9,10-cis-12,13-diepoxystearic + two monoepoxy impurities, 5. 12-hydroxyoleic, 6. threo-12,13-dihydroxyoleic, 7. threo-12,13-chlorohydroxyoleic + threo-13,12-chlorohydroxyoleic, 8. Artemisia absinthium acids and esters. Indicator: H_2SO_4; heat.

Acids or esters containing a primary hydroxyl group were separable from isomers with a secondary hydroxyl group. Acids and esters with an α-hydroxyl group were separable from isomers with the polar group in the middle of the chain. The following pairs of isomers were also separated: 9,10- and 12,13-epoxy-9-octadecanoic acids, 9-hydroxy-12-octadecanoic and 12-hydroxy-9-octadecanoic acids, 13-chloro-12-hydroxy-9-octadecenoic and 12-chloro-13-hydroxy-9-octadecenoic acids[194]. SUBBARAO et al.[195] separated numerous hydroxy, epoxy and halohydroxy acids including the cis- and trans-isomers of epoxydocosanoic acids.

Recently MORRIS[62] separated oxygenated fatty acids, in particular the threo and erythro isomers of dihydroxy acids, by chromatography on silica gel G impregnated with boric acid or a mixture of boric acid and silver nitrate.

MORRIS[66] separated numerous polyhydroxy acids on layers of silica gel G impregnated with boric acid, sodium borate or sodium arsenite, reagents which preferentially complex with threo glycols[196] (Fig. 4).

References p. 111 (authors) and p. 115 (apparatus etc.)

Partial glycerides

Mixtures of mono-, di-, and triglycerides are separated by chromatography on silica gel G, using mixtures of light petroleum and ether in the development. This is not a very satisfactory solvent system to separate a mixture of all three glycerides because of their widely different polarities. RYBICKA[119,120] used gradient elution to separate a mono-, di-, and triglyceride mixture when studying glycerolysis reactions. BROWN[166] recently separated the three glyceride types by a single development with a mixture of n-hexane/diethyl ether/acetic acid/methanol (90 : 20 : 2 : 3). Similar mixtures were separated by adsorbtion and reverse-phase chromatography[85]. Diglyceride mixtures were separated according to chain length by reverse-phase chromatography on silica gel G impregnated with undecane.

1- and 2-monoglyceride isomers were separated by RYBICKA[120] either by normal development with a light petroleum/diethyl ether mixture or by gradient elution with the same pair of solvents. PRIVETT AND BLANK[171] were unable to separate these isomers under practically the same conditions however. HOFMANN[101] recently resolved 1- and 2-monoglycerides on hydroxyl apatite. Hydroxyl apatite was prepared according to ANACKER AND STOY[196], passed through a 100-mesh sieve, and finally calcium sulphate added. The optimum conditions for the separation were obtained by developing the plate at 10° C with methylisobutyl ketone. The glycerides isomerised on the plate at higher temperatures.

1,2- and 1,3-diglycerides were separated[171] on silica gel G developed with a mixture of light petroleum/ether (60 : 40).

PRIVETT AND BLANK[172,173] described the quantitative analysis of mono-, di- and triglyceride mixtures by photodensitometry of the charred spots on the plate. Mixtures of saturated and unsaturated partial glycerides were quantitatively analysed by ozonizing the mixture reducing the separated ozonides and separating the glyceryl residues by thin-layer chromatography. 1- and 2-monoglyceride mixtures were oxidised with periodate before separation. The components were charred and the spots estimated by photodensitometry. In this way the four monoglyceride types and six of the seven diglyceride types were determined.

The alkoxy diglycerides of dogfish liver oil were characterised by MALINS[197] using thin-layer and gas-liquid chromatography.

The triglycerides of natural fats

The glyceride composition of natural fats has yet to be determined. Only in the last decade have we obtained an insight of the structure of natural fats by direct fractionation of the oil[254-257] or after oxidising the unsaturated fatty acids[258-261]. Perhaps the most useful information has been obtained by analysing the products of the partial hydrolysis of triglycerides by pancreatic lipase[262-266]. It is probable that thin-layer chromatography in conjunction with gas-liquid chromatography and enzymic hydrolysis will provide the complete answer.

Triglycerides cannot be separated by adsorption chromatography unless they differ widely in molecular weight *e.g.* tricaprin and tristearin, or unsaturation. Polar

Fig. 5. T.L.C. of seed oils containing epoxy, hydroxy, cyclopropene or keto groups. (Mangold and Morris, 1962.) Pet. ether/diethyl ether/acetic acid (70/30/2). Seed oils of: 1. Olea europa sativa, 2. Sterculia foetida, 3. Helianthus annuus, 4. Chrysanthemum coronarium, 5. Licania rigida, 6. Heliopsis pitcheriana, 7. Strophanthus kombe, 8. Strophanthus gratus, 9. Ricinus communis. Indicator: CrO_3/H_2SO_4; heat. Arrow shows position of plant sterols.

groups in the fatty-acid chains will obviously affect the polarity of the glyceride. MANGOLD AND MORRIS[194] examined numerous seed oils containing epoxy, hydroxy, keto, cyclopropene or acetylenic acids by silica-gel chromatography (Fig. 5). Extracts from healthy *Sterculia* nuts and from nuts which were diseased were compared.

Triglycerides were fractionated according to chain length and unsaturation by reverse-phase chromatography[83,85,87,89,90]. Critical pairs were separated after either hydrogenating or brominating the unsaturated glycerides on the plate. KAUFMANN AND DAS[89] resolved critical pairs of triglycerides, without prior modification, by multi-development reverse-phase chromatography. Kieselguhr G impregnated with liquid paraffin was the stationary phase and the plate was developed three times with a mixture of acetone/acetonitrile (7 : 4) saturated with paraffin (Fig. 6). Unsaturated glycerides, differing only in the configuration of the double bond, *e.g.* triolein and trielaidin were also separable.

MICHALEC[198] chromatographed mono-acid unsaturated triglycerides on silica gel G impregnated with paraffin oil.

The triglycerides of castor oil[198], corn oil[89], groundnut oil[89], human blood serum[198], lard[90], linseed oil[89,198], olive oil[90,198], sesame oil[198] and soya bean oil[89,198] were separated by reverse-phase chromatography. Recently the method was used in conjunction with gas-liquid chromatography and enzyme-hydrolysis to distinguish between cocoa butter and cocoa butter substitute[199].

References p. 111 (authors) and p. 115 (apparatus etc.)

Privett and Blank analysed triglyceride mixture after reductive ozonolysis[171–173]. Quantitative estimations of the GS_3, GS_2U, GSU_2 and GU_3 components were obtained by photodensitometry of the charred components. The glycerides of lard, corn oil, olive oil, cocoa butter and soya bean oil were quantitatively determined[173].

Barrett, Dallas and Padley[63,175,176] fractionated triglycerides according to their degree of unsaturation on a silica gel/silver nitrate adsorbent. Glycerides containing 0–10 double bonds were separated and also the mono-unsaturated isomers e.g. 1- and 2-oleodistearin. The plates were developed with a mixture of carbon tetrachloride/chloroform/acetic acid (60 : 40 : 0.5) containing small quantities of ethyl alcohol (0–2%) depending upon the degree of unsaturation of the glycerides (Fig. 7). A quantitative method based on photodensitometry of the charred components (50% aqueous phosphoric acid) was developed[175,176]. Cocoa butter, shea butter, malayan palm oil and lard triglycerides were determined quantitatively according to degree of unsaturation and also the mono-unsaturated isomers. The results compared favourably with the triglyceride composition determined by enzyme-hydrolysis. De Vries[65] separated on silica/silver nitrate triglycerides according to their degree of unsaturation and configuration of the double bond. Thus eleodistearin and elaidodistearin were resolved.

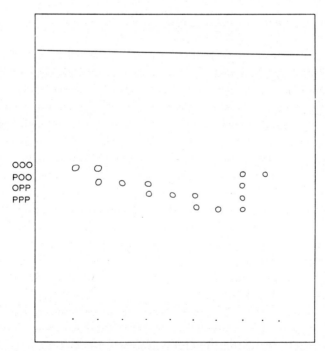

Fig. 6. Separation of critical pairs of triglycerides by multi-development reverse-phase chromatography (on kieselguhr impregnated with liquid paraffin developed 3 times with acetone/acetonitrile, 8 : 2) (Kaufmann and Das, 1962).

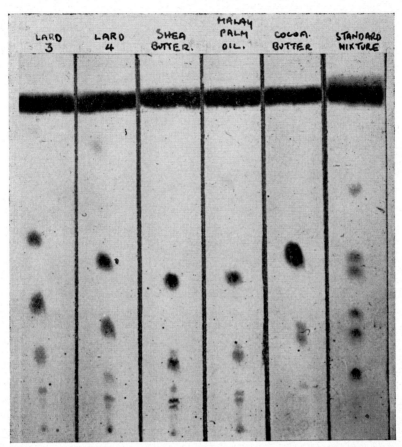

Fig. 7. T.L.C. of oils and fats on silica gel G impregnated with silver nitrate. (Barrett, Dallas and Padley, 1962.) Standard mixture (top to bottom) consists of tristearin, 2- and 1-oleodistearin, 2-linoleodistearin, oleodistearin and triolein. Each chromatogram was developed separately. Developing solvent: carbon tetrachloride (60 vol.), chloroform (40 vol.), acetic acid (0.5 vol.), ethyl-alcohol (0.4 vol.). Indicator: phosphoric acid and heat.

Phospholipids and sphingolipids

WEIKER[123] first separated serum lipids by thin-layer chromatography and the technique has since been widely used to fractionate phospholipids and sphingolipids.

JATZKEWITZ[187] separated brain phospholipids and in later work[97] estimated the components colorimetrically after eluting them from the plate. The solvent which gave the most satisfactory separations was a mixture of chloroform/methanol/water[187]. WAGNER[200] used thin-layer chromatography in conjunction with paper chromatography, counter-current distribution and infra-red spectroscopy to isolate and identify phospholipids (Fig. 8). Phosphatides extracted from various animal organs were also separated[201]. SKIDMORE AND ENTENMAN[202] fractionated rat liver phosphatides by two-dimensional chromatography. Characteristic hydrolysis products derived from phosphatidyl serine, phosphatidyl ethanolamine, phosphatidyl inositol, phosphatidyl choline, sphingomyelin and lysophosphatidyl choline were identified. SCHLEMMER[203]

References p. 111 (authors) and p. 115 (apparatus etc.)

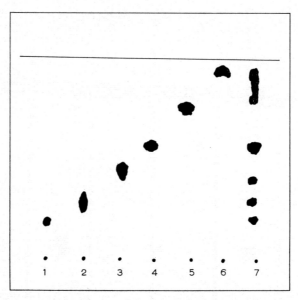

Fig. 8. Separation of phospholipids on silica gel; solvent: chloroform/methanol/water 65/25/4. 1. lysolecithin, 2. sphingomyelin, 3. lecithin, 4. cephalin, 5. cerebroside, 6. cardiolipid (Wagner, 1961).

separated phosphatides and MANGOLD AND KAMMERECK[58] obtained sharp separations of phospholipids (lecithins, cephalines and sphingomyelins) on silica gel G containing ammonium sulphate (10%). The structure of lecithins[204,205,173] was determined by reductive ozonolysis and reverse-phase chromatography of the reduction products on siliconised plates.

Serum phospholipids have been separated by numerous workers[39,51,60,88,162, 200-203,207]. SKIPSKI et al.[60] separated phosphatidyl serine from other phosphatides. Serum phospholipids have been separated on layers of unbound alumina and HABERMANN[163] fractionated phospholipids on silica gel G, ashed the separated components, and determined the residual phosphorus colorimetrically.

The chromatography of the gangliosides e.g. from beef brain, has been described[97, 208-210,248] and the method used to isolate four crystalline gangliosides[210]. HONEGGER separated into thirty two components the brain lipids from multiple sclerosis patients; the extracted components were estimated semi-quantitatively and qualitatively[206,207]. KOCHETKOV et al.[211,212] chromatographed cerebrosides and sphingosine derivatives and FUJINO et al.[213] separated sphingosine bases. Phytosphingosine, *erythro*-sphingosine, *threo*-sphingosine, dihydrosphingosine and 3-O-methy sphingosine were recently chromatographed on silica gel G developed with a chloroform/methanol/ammonia (2 N) mixture (40 : 10 : 1) by SAMBASIVARAO AND MCCLUER[214].

Autoxidation and antioxidants

The detection by thin-layer chromatography of hydroperoxides formed during the autoxidation of methyllinoleate was described by PRIVETT AND BLANK[215]. They have also investigated the initial stages of autoxidation and were able to isolate by thin-layer chromatography non-hydroperoxide substances formed during the induction period[184]. This material exhibited a broad adsorption band in the conjugated diene adsorption region of the U.V. spectrum. Organic peroxides were separated by STAHL[34], and MARUYAMA AND GOTO[216]. Ketodiens formed during the oxidation of methyl linoleate by lipoxidase were isolated and characterised by column and thin-layer chromatography[217].

KNAPPE AND PETERI[218] identified alkyl hydroperoxides, dialkyl peroxides, diacyl peroxides and hydroperoxides of ketones by thin-layer chromatography on silica gel G developed with mixtures of either carbontetrachloride/toluene or acetic acid/toluene. MARCUSE[219] separated alkanals and alkanones for studies in fat oxidation and flavour. A number of acetals and their corresponding aldehydes were chromatographed by KORE et al.[220].

Natural and synthetic antioxidants were analysed on layers of silica containing oxalic acid by SEHER[131,153]. Tocopherol mixtures (α, β, γ) were also separated on silica or alumina and the separated components estimated semi-quantitatively by measuring the spot areas[154]. BARBIER[221] separated naturally occurring quinones.

MEYER[222] described methods of determining antioxidants in fats and separated B.H.A., B.H.T. and nordihydroguaiaretic acid on silica gel G/kieselguhr. GÄNSHIRT AND MORIANZ[223] determined p-hydroxybenzoate quantitatively and recently DAVIDEK[224,225] detected antioxidants in fats by chromatography on polyamide powder, including propylgallate, B.H.A., nordihydroguaiaretic acid and gallic acid and its esters. The plates were developed at an angle of 20–30° with a chloroform/ethanol or methanol (7 : 3) mixture.

Fat-soluble vitamins and carotenoids

Carotenoid aldehydes extracted from a number of vegetables and fruit were separated by WINTERSTEIN et al.[59,227] using adsorption and reverse-phase chromatography. A sensitive method of detecting faintly coloured aldehydes ($\sim$0.03γ) by conversion to deeply coloured rhodanine derivatives was described[227]. GROB AND PFLUGSHAUPT (ref. 228) separated carotenoids and xanthophyll degradation products by developing chromatograms on silica gel G with a benzene/light petroleum/ethanol (50 : 50 : 2) mixture.

DAVIDEK AND BLATTNA[38] separated practically all the fat-soluble vitamins (S, D_2, E, K_1, K_2, K_3, and α- and β-carotene) on layers of powdered alumina. Colour reactions of the vitamins with perchloric acid (70%) and sulphuric acid (20%) were described. Vitamin A esters in various fish oils were detected by MANGOLD AND MALINS[186] and vitamin A and vitamin A_2 isomers were separated on silica gel developed with light petroleum (40–60°)/methyl heptenone (11 : 2) by PLANTA et al.[229].

The vitamins of group B (B_1, B_2, B_6, nicotinamide, biotine, and calcium pantothenate) and vitamin C were chromatographed on silica gel G with benzene/methanol/acetone/acetic acid (70 : 20 : 5 : 5), by GÄNSHIRT AND MALZACHER[226].

Tocopherols were determined semi-quantitatively[154] by measuring the spot areas of the separated components. Vitamin D_2 and various sterols were resolved on kieselguhr by COPIUS PEEREBOOM AND BEEKES[230].

Ubiquinones were isolated by adsorption chromatography from crude lipid extracts and separated by reverse-phase chromatography[231]. WAGNER AND DENGLER[164] also quantitatively estimated ubiquinones present in the acetone extracts of heart muscle, pig retinas and bakers yeast. Ubiquinones, vitamin K, tocopherol, phosphatides, cholesterol and cholesterol esters were separated by thin-layer chromatography and the fractions eluted and estimated. MARTIUS et al.[232,233] chromatographed the K vitamins when studying their metabolism.

Minor constituents of fats

The unsaponifiable fractions obtained from refined soya bean oil were examined by HOFFMANN et al.[234] when studying the effects of unsaponifiable materials on the oxidative stability of oils. PRIVETT et al.[235] chromatographed various extracts of parsley seed oil including unsaponifiable compounds and developed a solvent extraction method to isolate myristicin (5-allyl-1-methoxy-2,3-methylenedioxybenzene) and petroselenic acid. Acetylenic alcohols and glycols were separated on thin layers of alumina by AHKREM et al.[236].

The toxic material produced by the action of a strain of the fungus Aspergillus Flavus[241] was separated and isolated by thin-layer chromatography[237-240]. GRACIAN AND MARTEL[244] isolated a fluorescent material from sulphur olive oil.

The detection of non-lipid materials in edible products is increasing in importance with the wider use of insecticides in farming and plastics in packaging. BÄUMLER AND RIPPSTEIN[244] detected insecticides by thin-layer chromatography and YAMANURA et al.[245] separated chlorinated pesticides. STANSBY[246] discussed the potential applications of thin-layer chromatography to analyse foods. Food-colouring agents including vitamin A were identified by thin-layer chromatography[141,247]. COPIUS PEEREBOOM chromatographed plasticizers and identified them using specific colour reactions.

11. CONCLUSION

In this review the separation of lipids, excluding steroids, by thin-layer chromatography is discussed. Chemists and biochemists are widely applying this invaluable technique because sharp separations are obtained in a relatively short time. Small amounts of oil and synthetic products are readily characterised[172,249] and reactions [250-252] or chromatographic columns[253] monitored by thin-layer chromatography.

Quantitative methods of determining numerous classes of compounds have been described and because many separations are unique to thin-layer chromatography the method has been used for preparative work.

Closely related compounds and some isomers are separated by chromatography on modified layers of silica and by multideveloping the chromatoplate. Further research along these lines is required before some of the more similar isomers are separated.

REFERENCES

1. N. A. IZMAILOV AND M. S. SCHRAIBER, *Farmatsiya No. 3*, (1938) 1.
2. T. I. WILLIAMS, *Introduction to Chromatography*, Blackie & Son, Glasgow, 1947.
3. J. E. MEINHARD AND N. F. HALL, *Anal. Chem.*, 21 (1949) 185.
4. J. G. KIRCHNER, J. M. MILLER AND G. E. KELLER, *Anal. Chem.*, 23 (1951) 420.
5. R. H. REITSEMA, *Anal. Chem.*, 26 (1954) 960.
6. E. STAHL, *Pharmazie*, 11 (1956) 633.
7. E. DEMOLE, *J. Chromatog.*, 1 (1958) 24.
8. E. DEMOLE, *J. Chromatog.*, 6 (1961) 2.
9. E. DEMOLE, *Chromatog. Rev.*, Vol. 4, Ed. by M. LEDERER, Elsevier Publishing Company, Amsterdam.
10. K. FONTELL, R. T. HOLMAN AND G. LAMBERTSON, *J. Lipid Res.*, 1 (1960) 482.
11. A. JENSEN, *Tidsskri. Kjemi, Bergvesen Met.*, 21 (1961) 14.
12. H. K. MANGOLD, *J. Am. Oil Chemists' Soc.*, 38 (1961) 702.
13. C. MICHALEC, *Chem. Listy*, 55 (1961) 953.
14. L. J. MORRIS, *Chromatography*, Ed. by HEFTMANN, Reinhold Publishers, N.Y.
15. G. NEUBERT, *Chem. Labor. Betrieb*, 11 (1960) 23.
16. J. W. COPIUS PEEREBOOM, *J. Am. Oil Chemists' Soc.*, 39 (1962) 22.
17. Z. PROCHAZKA, *Chem. Listy*, 55 (1961) 974.
18. E. STAHL, *Angew. Chem.*, 73, No. 19, (1961) 646.
19. E. STAHL, *Z. Anal. Chem.*, 181 (1961) 303.
20. E. VIOQUE, *Grasas Aceites (Seville, Spain)*, 11 (1960) 223.
21. H. WAGNER, *Mitt. Gebiete Lebensm. Hyg.*, 51 (1960) 416.
22. E. G. WOLLISH, M. SCHMALL AND M. HAWRYLYSHYN, *Anal. Chem.*, 33 (1961) 1138.
23. E. G. WOLLISH, *Mickrochem. J., Symp. Ser.*, 2 (1962) 687.
24. K. RANDERATH, *Dünnschicht-Chromatographie*, Verlag Chemie, Weinheim, 1962.
25. E. V. TRUTER, *Thin-Film Chromatography*, Cleaver Hume Press Ltd., London, 1963.
26. E. STAHL, in conjunction with other workers, *Thin-Layer Chromatography*, Springer-Verlag, Berlin, Göttingen, Heidelberg.
27. J. G. KIRCHNER AND G. L. KELLER, *J. Am. Chem. Soc.*, 72 (1950) 1867.
28. J. M. MILLER AND J. G. KIRCHNER, *Anal. Chem.*, 26 (1954) 2002.
29. J. G. KIRCHNER AND J. M. MILLER, *Ind. Eng. Chem.*, 44 (1952) 318.
30. J. G. KIRCHNER AND J. M. MILLER, *J. Agr. Food Chem.*, 1 (1953) 512.
31. J. G. KIRCHNER, J. M. MILLER AND R. G. RICE, *J. Agr. Food Chem.*, 2 (1954) 1031.
32. J. G. KIRCHNER AND J. M. MILLER, *J. Agr. Food Chem.*, 5 (1957) 283.
33. R. H. REITSEMA, *J. Am. Pharm. Assoc., Sci. Ed.*, 43 (1954) 414.
34. E. STAHL, *Chemiker Ztg.*, 82 (1958) 323.
35. E. STAHL, *Parfuem. Kosmetik*, 39 (1958) 564.
36. K. MUTTER AND J. F. HOFFSTETTER, *Anal. Chem.*, 33 (1961) 1138.
37. I. BEKERSKY, *Anal. Chem.*, 35 (1963) 261.
38. J. DAVIDEK AND J. BLATTNA, *J. Chromatog.*, 7 (1962) 204.
39. A. VOCIKOVA, V. FELT AND J. MALIKOVA, *J. Chromatog.*, 9 (1962) 301.
40. P. M. BOLL, *Chem. Anal.*, 51 (1962) 52.
41. G. R. DUNCAN, *J. Chromatog.*, 8 (1962) 37.
42. T. M. LEES AND P. J. DE MURIA, *J. Chromatog.*, 8 (1962) 108.
43. M. BARBIER, H. JAGER, H. TOBIAS AND E. WYSS, *Helv. Chim. Acta*, 42 (1959) 2440.
44. G. MACHATA, *Mikrochim. Acta*, (1960) 79.
45. F. G. RITTER AND G. M. MEYER, *Nature*, 193 (1962) 941.
46. M. MOTTIER AND M. POTTERAT, *Anal. Chim. Acta*, 13 (1955) 46.
47. A. GAMP, *Experientia*, 18 (1962) 292.
48. J. J. PEIFER, *Mikrochim. Acta*, (1962) 529.
49. A. F. HOFMANN, *Anal. Biochem.*, 3 (1962) 145.
50. R. WASIKY, *Anal. Chem.*, 34 (1962) 1346.
51. K. B. LIE AND J. F. MYE, *J. Chromatog.*, 8 (1962) 75.
52. P. ROSI AND P. HAMILTON, *J. Chromatog.*, 9 (1962) 388.
53. J. J. WREN, *J. Chromatog.*, 4 (1960) 176.
54. P. D. KLEIN, *Anal. Chem.*, 34 (1962) 733.
55. R. HERNANDEZ, R. HERNANDEZ, JR. AND L. R. AXELROD, *Anal. Chem.*, 33 (1961) 370.
56. E. STAHL, *Arch. Pharm.*, 292 (1959) 411.
57. J. J. PEIFER AND R. A. MUESING, *J. Am. Oil Chemists' Soc.*, 38 (1961) 702.
58. H. K. MANGOLD AND R. KAMMERECK, *J. Am. Oil Chemists' Soc.*, 38 (1961) 201; 708.
59. A. WINTERSTEIN, A. STUDER AND R. RÜEGG, *Chem. Ber.*, 93 (1960) 2951.
60. V. P. SKIPSKI, R. F. PETERSON AND M. BARCLAY, *J. Lipid Res.*, 3 (1962) 467.

61. J. Halmekoski, *Suomen Kemistilehti*, 35 B(1962) 39.
62. L. J. Morris, *Chem. Ind. (London)*, (1962) 1238.
63. C. B. Barrett, M. S. J. Dallas and F. B. Padley, *Chem. Ind.*, (1962) 1050.
64. B. de Vries, *Chem. Ind.*, (1962) 1049.
65. B. de Vries, *Fall Meeting of the Am. Oil Chemists' Soc.*, Toronto, Canada, Oct. 1962.
66. L. J. Morris, private communication.
67. R. L. Stedman, A. P. Swain and W. Rusaniwsky, *J. Chromatog.*, 4 (1960) 252.
68. E. Lederer and M. Lederer, *Chromatography*, 2nd. Ed. 1957, p. 61. Elsevier Publishing Co., Amsterdam.
69. B. Borgström, *Acta Physiol. Scand.*, 25 (1952) 101.
70. W. Trappe, *Biochem. Z.*, 306 (1940) 316.
71. W. Trappe, *Biochem. Z.*, 307 (1941) 97.
72. W. Trappe, *Z. Physiol. Chem.*, 273 (1942) 177.
73. H. Brockmann and H. Schodder, *Ber.*, 74 (1941) 73.
74. E. Stahl, *Chemiker Ztg.*, 85 (1961) 371.
75. K. Teichert, E. Mutschlet and H. Röchelmeyer, *Deut. Apotheker Ztg.*, 100 (1960) 477.
76. P. Wollenweber, *J. Chromatog.*, 9 (1962) 369.
77. H. Wagner, *Fette Seifen Anstrichmittel*, 63 (1961) 1119.
78. Th. Wieland, G. Lüben and H. Determann, *Experientia*, 18 (1962) 432.
79. G. M. Badger, J. K. Donnelly and T. M. Spotswood, *J. Chromatog.*, 10 (1963) 397.
80. H. Determann, *Experientia*, 18 (1962) 431.
81. Th. Wieland and H. Determann, *Experientia*, 18 (1962) 432.
82. D. C. Malins and H. K. Mangold, *J. Am. Oil Chemists' Soc.*, 37 (1960) 576.
83. H. P. Kaufmann, Z. Makus and B. Das, *Fette Seifen Anstrichmittel*, 63 (1961) 807.
84. H. P. Kaufmann and T. H. Khoe, *Fette Seifen Anstrichmittel*, 64 (1962) 81.
85. H. P. Kaufmann and Z. Makus, *Fette Seifen Anstrichmittel*, 62 (1960) 1014.
86. H. P. Kaufmann and Suko Young, *Fette Seifen Anstrichmittel*, 63 (1961) 828.
87. H. P. Kaufmann, Z. Makus and T. H. Khoe, *Fette Seifen Anstrichmittel*, 63 (1961) 689.
88. H. P. Kaufmann, Z. Makus and F. Deicke, *Fette Seifen Anstrichmittel*, 63 (1961) 235.
89. H. P. Kaufmann and B. Das, *Fette Seifen Anstrichmittel*, 64 (1962) 214.
90. H. P. Kaufmann, Z. Makus and T. H. Khoe, *Fette Seifen Anstrichmittel*, 64 (1962) 1.
91. O. Wiss and V. Gloor, *Z. Physiol. Chem.*, 310 (1958) 260.
92. E. Knappe and D. Peteri, *Z. Anal. Chem.*, 188 (1962) 184.
93. E. Stahl, *Z. Anal. Chem.*, 181 (1961) 303.
94. L. Birkofer, C. Kaiser, H. A. Meyer-Stoll and F. Suppan, *Z. Naturforsch.*, 17B (1962) 352.
95. J. Davidek and Z. Prochazka, *Collection Czech. Chem. Commun.*, 26 (1961) 2947.
96. K. Randerath, *Angew. Chem.*, 74 (1962) 484.
97. H. Jatzkewitz, *Z. Physiol. Chem.*, 326 (1961) 61.
98. G. Weimann and K. Randerath, *Experientia*, 19 (1963) 49.
99. E. Stahl and U. Kaltenbach, *J. Chromatog.*, 5 (1961) 351.
100. T. F. Brodasky, *Anal. Chem.*, in press.
101. A. F. Hofmann, *J. Lipid Res.*, 3 (1961) 391.
102. K. K. Caroll, *J. Lipid Res.*, 2 (1961) 135.
103. J. Hirsch, *J. Lipid Res.*, 4 (1963) 1.
104. H. Brockmann and F. Volpers, *Chem. Ber.*, 8 (1949) 95.
105. E. Stahl, *Pharm. Rundschau*, 2 (1959) 1.
106. H. S. Knight and S. Groennings, *Anal. Chem.*, 26 (1954) 1549.
107. H. H. Strain, *Chromatographic Adsorption Analysis*, Interscience Publishers, Inc., New York, N.Y., 1942.
108. M. Brenner and G. Pataki, *Helv. Chim. Acta*, 44 (1961) 1420.
109. V. Cerny, J. Joska and L. Labler, *Collection Czech. Chem. Commun.*, 26 (1961) 1658.
110. M. Brenner and A. Niederwieser, *Experientia*, 17 (1961) 237.
111. W. L. Stanley, R. M. Ikeda and S. Cook, *Food Technol.*, 15 (1961) 381.
112. S. Hermanek, V. Schwartz and Z. Cekan, *Collection Czech. Chem. Commun.*, 26 (1961) 1669.
113. H. Logoni and A. Wortmann, *Milchwissenschaft*, 11 (1956) 206.
114. M. Mottier, *Mitt. Gebiete Lebensm. Hyg.*, 43 (1952) 118.
115. E. A. Mistryukov, *J. Chromatog.*, 9 (1962) 311.
116. E. A. Mistryukov, *J. Chromatog.*, 9 (1962) 314.
117. M. Chakrabarty, *J. Am. Oil Chemists' Soc.*, 38 (1961) 715.
118. G. A. Dhopeshwarker and J. F. Mead, *J. Am. Oil Chemists' Soc.*, 38 (1961) 297.
119. S. M. Rybicka, *Chem. Ind. (London)*, (1962) 308.
120. S. M. Rybicka, *Chem. Ind. (London)*, (1962) 1947.
121. W. Matthias, *Naturwissenschaften*, 41 (1954) 18.

122. Y. FURUKAWA, *Nippon Kagaku Zasshi*, 80 (1959) 45.
123. H. WEIKER, *Klin. Wochschr.*, 37 (1959) 763.
124. K. V. GIRI, *Nature*, 173 (1954) 1194.
125. J. A. THOMA, *Anal. Chem.*, 35 (1963) 214.
126. F. SNYDER AND N. STEPHENS, *Anal. Biochem.*, 4 (1962) 128.
127. J. F. HERNDON, H. E. APPERT, J. C. TOUCHSTONE AND C. N. DAVIS, *Anal. Chem.*, 34 (1962) 1061.
128. G. PASTUSKA AND H. TRINKS, *Chemiker Ztg.*, 86 (1962) 135.
129. H. A. RAMSEY, *Anal. Biochem.*, 5 (1963) 83.
130. J. F. HERNDON, H. E. APPERT, J. C. TOUCHSTONE AND C. N. DAVID, *Anal. Chem.*, 34 (1962) 1061.
131. A. SEHER, *Fette Seifen Anstrichmittel*, 61 (1959) 345.
132. M. BRENNER, A. NIEDERWIESER, G. PATAKI AND A. R. FAHRMY, *Experientia*, 18 (1962) 101.
133. H. K. MANGOLD, J. L. GELLERMAN AND H. SHLENK, *Federation Proc.*, 17 (1958) 268.
134. H. K. MANGOLD, *Fette Seifen Anstrichmittel*, 61 (1959) 877.
135. E. JANTZEN AND H. ANDREAS, *Chem. Ber.*, 92 (1959) 1427.
136. H. K. MANGOLD AND R. KAMMERECK, *Chem. Ind. (London)*, (1961) 1032.
137. H. K. MANGOLD, R. KAMMERECK AND D. C. MALINS, *Proceedings 1961, International Symposium on Microchemical Techniques* (Microchem. J., Symposium II).
138. W. E. BAUMGARTNER, L. S. LAZER, A. M. DALZIEL, E. V. CARDINAL AND E. L. VARNER, *J. Agr. Food Chem.*, 7 (1959) 422.
139. L. LEITCH, P. E. GAGNON AND A. CAMBRON, *Can. J. Res.*, 28 B(1950) 256.
140. J. M. MILLER AND J. G. KIRCHNER, *Anal. Chem.*, 25 (1953) 1107.
141. E. STAHL, *Arch. Pharm.*, 293 (1960) 531.
142. I. M. HAIS AND K. MACEK, *Handbuch der Papier Chromatographie*, Verlag Gustav Fischer, Jena, 1958.
143. F. FEIGL, *Spot Tests in Organic Analysis*, 6th ed., Elsevier Publishing Co., Amsterdam, 1960.
144. R. P. A. TIMS AND J. A. G. LAROSE, *J. Am. Oil Chemists' Soc.*, 39 (1962) 232.
145. J. BARROLLIER, *Naturwissenschaften*, 48 (1961) 404.
146. W. LICHTENBERGER, *Z. Anal. Chem.*, 185 (1961) 101.
147. F. EISENBERG, *J. Chromatog.*, 9 (1962) 390.
148. E. BÜRKI AND H. BOLLINGER, *J. Am. Oil Chemists' Soc.*, 38 (1961) 717.
149. D. ABELSON, *Nature*, 188 (1960) 850.
150. O. S. PRIVETT, private communication.
151. E. HANSBURY, J. LANGHAM AND G. O. DONALD, *J. Chromatog.*, 9 (1962) 393.
152. L. J. MORRIS, R. T. HOLMAN AND K. FONTELL, *J. Am. Oil Chemists' Soc.*, 37 (1960) 323.
153. A. SEHER, *Nahrung*, 4 (1960) 466.
154. A. SEHER, *Mikrochim. Acta*, (1961) 308.
155. F. W. HEFENDEL, *Planta Med.*, 8 (1960) 65.
156. H. R. GETZ AND D. D. LAWSON, *J. Chromatog.*, 7 (1962) 266.
157. J. HILTON AND W. B. HALL, *J. Chromatog.*, 7 (1962) 266.
158. R. KAMMERECK, *J. Am. Oil Chemists' Soc.*, 38 (1961) 717.
159. J. A. WILLIAMS, A. SHARMA, L. J. MORRIS AND R. T. HOLMAN, *Proc. Soc. Exptl. Biol. Med.*, 105 (1960) 192.
160. J. DAVIDEK AND E. DAVIDKOVA, *Pharmazie*, 16 (1961) 352.
161. K. HUHNSTOCK AND H. WEIKER, *Klin. Wochschr.*, 38 (1960) 1249.
162. H. GÄNSHIRT, F. W. KOSS AND K. MORIANZ, *Arzneimittel-Forsch.*, 10 (1960) 943.
163. E. HABERMANN, G. BANDTLOW AND B. KRUSCHE, *Klin. Wochschr.*, 39 (1961) 816.
164. H. WAGNER AND B. DENGLER, *Biochem. Z.*, 336 (1962) 380.
165. E. VIOQUE AND R. T. HOLMAN, *J. Am. Oil Chemists' Soc.*, 39 (1962) 63.
166. J. L. BROWN AND J. M. JOHNSTON, *J. Lipid Res.*, 3 (1962) 480.
167. K. HEYNS AND H. F. GRUETZMACHER, *Angew. Chem.*, 74 (1962) 387.
168. R. B. FISHER, D. S. PARSONS AND G. A. MORRISON, *Nature*, 161 (1948) 764.
169. M. BRENNER AND A. NIEDERWIESER, *Experientia*, 16 (1960) 378.
170. S. JEAN PURDY AND E. V. TRUTER, *Chem. Ind. (London)*, (1962) 506.
171. O. S. PRIVETT AND M. L. BLANK, *J. Lipid Res.*, 2 (1961) 37.
172. O. S. PRIVETT, M. L. BLANK AND W. O. LUNDBERG, *J. Am. Oil Chemists' Soc.*, 38 (1961) 312.
173. O. S. PRIVETT AND M. L. BLANK, *J. Am. Oil Chemists' Soc.*, 40 (1963) 170.
174. O. S. PRIVETT AND M. L. BLANK, *J. Am. Chem. Soc.*, 39 (1962) 520.
175. C. B. BARRETT, M. S. J. DALLAS AND F. B. PADLEY, *36th Fall Meeting, Am. Oil Chemists' Soc.*, October 1962.
176. C. B. BARRETT, M. S. J. DALLAS AND F. B. PADLEY, *J. Am. Oil Chemists' Soc.*, in press.
177. A. S. CSALLANY AND H. H. DRAPER, *Anal. Biochem.*, 4 (1962) 418.
178. C. G. HONEGGER, *Helv. Chim. Acta*, 45 (1962) 1409.

179. M. E. MORGAN, *J. Chromatog.*, 9 (1962) 379.
180. S. H. ENER, *J. Am. Oil Chemists' Soc.*, 38 (1961) 717.
181. B. BAEHLER, *Helv. Chim. Acta*, 45 (1962) 309.
182. R. G. JENSEN AND T. SAMPUGNA, *J. Dairy Sci.*, (1962) 435.
183. S. RUGGIERI, *Nature*, 193 (1962) 1282.
184. O. S. PRIVETT AND M. L. BLANK, *J. Am. Oil Chemists' Soc.*, 39 (1962) 465.
185. H. DAHN AND H. FUCHS, *Helv. Chim. Acta*, 45 (1962) 261.
186. H. K. MANGOLD AND D. C. MALINS, *J. Am. Oil Chemists' Soc.*, 37 (1960) 383.
187. H. JATZKEWITZ AND E. MEHL, *Z. Physiol. Chem.*, 320 (1960) 251.
188. W. C. VOGEL, W. M. DOIZAKI AND L. ZIEVE, *J. Lipid Res.*, 3 (1962) 138.
189. H. K. MANGOLD AND N. TUNA, *Federation Proc.*, 20 (1961) 268.
190. A. J. DURBETAKI, *Anal. Chem.*, 28 (1956) 2000.
191. L. J. MORRIS, R. T. HOLMAN AND K. FONTELL, *J. Lipid Res.*, 2 (1961) 68.
192. E. VIOQUE, L. J. MORRIS AND R. T. HOLMAN, *J. Am. Oil Chemists' Soc.*, 38 (1961) 485, 489.
193. L. J. MORRIS, H. HAYES AND R. T. HOLMAN, *J. Am. Oil Chemists' Soc.*, 38 (1960) 316.
194. H. K. MANGOLD AND L. J. MORRIS, *Vth I.S.F. Congress*, London, 1962.
195. R. SUBBARAO, M. W. ROOMI, M. R. SUBBARAM AND K. T. ACHAYA, *J. Chromatog.*, 9 (1962) 295.
196. W. F. ANACKER AND V. STOY, *Biochem. Z.*, 330 (1958) 141.
197. D. C. MALINS, *Chem. Ind. (London)*, (1960) 1359.
198. C. MICHALEC, M. SULC AND J. MESTAN, *Nature*, 193 (1962) 63.
199. H. P. KAUFMANN, H. WESSELS AND B. DAS, *Fette Seifen Anstrichmittel*, 64 (1962) 723.
200. H. WAGNER, *Fette Seifen Anstrichmittel*, 62 (1960) 1115.
201. H. WAGNER, L. HORHÄMMER AND P. WOLFF, *Biochem. Z.*, 334 (1961) 175.
202. W. D. SKIDMORE AND C. ENTENMAN, *J. Lipid Res.*, 3 (1962) 471.
203. W. SCHLEMMER, *Boll. Soc. Ital. Biol. Sper.*, 37 (1961) 134.
204. O. S. PRIVETT AND M. L. BLANK, *Vth I.S.F. Congress*, London, 1962.
205. L. M. BLANK, O. S. PRIVETT AND J. A. SCHMIT, *J. Food Sci.*, 27 (1962) 463.
206. C. G. HONEGGER, *Helv. Chim. Acta*, 45 (1962) 281.
207. C. G. HONEGGER, *Helv. Chim. Acta*, 45 (1962) 2020.
208. J. A. DAIN, H. WEICKER, G. SCHMIDT AND S. J. THANNHAUSER, *Symposium on Sphingo-Lipidoses and Allied Diseases*, 1961.
209. E. KLENK AND W. GIELEN, *Z. Physiol. Chem.*, 323 (1961) 126.
210. R. KUHN, H. WIEGANDT AND H. EGGE, *Angew. Chem.*, 73 (1961) 580.
211. N. K. KOCHETKOV, I. G. ZHUKOVA AND I. S. GLUKHODED, *Doklady Akad. Nauk SSSR*, 139 (1961) 608.
212. N. K. KOCHETKOW, I. G. ZHUKOVA AND I. S. GLUKHODED, *Biochim. Biophys. Acta*, 60 (1962) 431.
213. Y. FUJINO AND I. ZABIN, *J. Biol. Chem.*, 237 (1962) 2069.
214. K. SAMBASIVARAO AND R. H. MCCLUER, *J. Lipid Res.*, 4 (1963) 106.
215. O. S. PRIVETT, *Proceedings Flavour Chemistry Symposium*, Campbell Soup Coy., Camden, N.J., U.S.A., 1961, p. 147.
216. K. MARUYAMA, K. OMOE AND R. GOTO, *J. Chem. Soc. Japan*, 77 (1956) 1496.
217. E. VIOQUE AND R. T. HOLMAN, *Arch. Biochem. Biophys.*, 99 (1962) 522.
218. E. KNAPPE AND D. PETERI, *Z. Anal. Chem.*, 190 (1962) 386.
219. R. MARCUSE, *J. Chromatog.*, 7 (1962) 407.
220. S. A. KORE, E. I. SHEPELENKOVA AND E. M. CHERNOVA, *Maslob-Zhir. Prom.*, 28 (1962) 32.
221. M. BARBIER, *J. Chromatog.*, 2 (1958) 649.
222. H. MEYER, *Deut. Lebensm. Rundschau*, 57 (1961) 170.
223. H. GÄNSHIRT AND K. MORIANZ, *Arch. Pharm.*, 293 (1960) 1065.
224. J. DAVIDEK AND J. POKORNY, *Z. Lebensm. Untersuch.-Forsch.*, 115 (1961) 113.
225. J. DAVIDEK, *J. Chromatog.*, 9 (1962) 363.
226. H. GÄNSHIRT AND A. MALZACHER, *Naturwiss.*, 47 (1960) 279.
227. A. WINTERSTEIN AND B. HEGEDUS, *Chimia (Aarau)*, 14 (1960) 18.
228. E. C. GROB AND R. P. PFLUGSHAUPT, *Helv. Chim. Acta*, 45 (1962) 1592.
229. C. V. PLANTA, U. SCHWIETER, L. CHOPARD-DIT-JEAN, R. RÜEGG, M. KOFLER AND O. ISLER, *Helv. Chim. Acta*, 45 (1962) 548.
230. J. W. COPIUS PEEREBOOM AND H. W. BEEKES, *J. Chromatog.*, 9 (1962) 316.
231. H. WAGNER, L. HOERHAMMER AND B. DENGLER, *J. Chromatog.*, 7 (1962) 211.
232. M. BILLETER AND C. MARTIUS, *Biochem. Z.*, 334 (1961) 304.
233. W. STOFFEL AND C. MARTIUS, *Biochem. Z.*, 333 (1960) 440.
234. L. R. HOFMANN, H. A. MOSER, C. D. EVANS AND J. C. COWAN, *J. Am. Oil Chemists' Soc.*, 39 (1962) 323.
235. O. S. PRIVETT, J. D. NADENICEK, R. P. WEBER AND F. J. PUSCH, in press.

236. A. A. AHKREM, A. I. KUZNETSOVA, YU. A. TITOV AND I. S. LEVINA, *Izv. Akad. Nauk SSSR, Otd. Khim. Nauk.*, (1962) 657.
237. H. DE IONGH, R. K. BEERTHUIS, R. O. BLAS, C. B. BARRETT AND W. O. ORD, *Biochim. Biophys. Acta*, in press.
238. A. S. M. VAN DER ZIGDEN, B. W. A. A. KOELENSMID, C. B. BARRETT, W. O. ORD AND J. PHILP, *Nature*, 195 (1962) 1060.
239. T. J. COOMES, J. A. CORNELIUS AND G. SHON, *Chem. Ind. (London)*, (1963) 367.
240. B. NESBITT, J. O'KELLY, K. SARGENT AND A. SHERIDAN, *Nature*, 195 (1962) 1062.
241. K. SARGENT, A. SHERIDAN, J. O'KELLY AND R. A. B. CARNAGHAN, *Nature*, 193 (1961) 1096.
242. J. GRACIAN AND J. MARTEL, *Grasas Aceites (Seville, Spain)*, 13 (1962) 128.
243. J. W. COPIUS PEEREBOOM, *J. Chromatog.*, 4 (1960) 323.
244. J. BÄUMLER AND S. RIPPSTEIN, *Helv. Chim. Acta*, 44 (1961) 1162.
245. J. YAMANURA AND T. NIWAGUCHI, *Kagaku Keisatsu Kenkyusho Hokoku*, 13 (1960) 450.
246. M. E. STANSBY, *Food Technol.*, 15 (1961) 378.
247. J. DAVIDEK, J. POKORNY AND G. JANICEK, *Z. Lebensm. Untersuch.-Forsch.*, 116 (1961) 13.
248. R. KUHN AND H. WIEGANDT, *Chem. Ber.*, 96 (1963) 866.
249. G. B. CRUMP, *Nature*, 193 (1962) 674.
250. E. H. GRUGER, D. C. MALINS AND E. J. GOUHTZ, *J. Am. Oil Chemists' Soc.*, 37 (1960) 214.
251. O. S. PRIVETT AND C. NICKELL, *J. Am. Oil Chemists' Soc.*, 39 (1962) 414.
252. *Chem. Eng. News*, 39 (1961) 42.
253. T. H. APPLEWHITE, M. J. DIAMOND AND L. A. GOLDBLATT, *J. Am. Oil Chemists' Soc.*, 38 (1961) 609.
254. H. J. DUTTON AND J. A. CANNON, *J. Am. Oil Chemists' Soc.*, 33 (1956) 46.
255. C. R. SCHOFIELD AND H. J. DUTTON, *J. Am. Oil Chemists' Soc.*, 36 (1959) 325.
256. O. Y. QUIMBY, R. L. WILLE AND E. S. LUTTON, *J. Am. Oil Chemists' Soc.*, 30 (1953) 186.
257. R. W. RIEMENSCHNEIDER, *J. Am. Oil Chemists' Soc.*, 31 (1954) 266.
258. T. P. HILDITCH AND C. H. LEA, *J. Chem. Soc.*, (1927) 3106.
259. A. R. S. KARTHA, *J. Am. Oil Chemists' Soc.*, 30 (1953) 280.
260. A. R. S. KARTHA, *J. Am. Oil Chemists' Soc.*, 30 (1953) 326.
261. C. G. YOUNGS, *J. Am. Oil Chemists' Soc.*, 38 (1961) 62.
262. P. SAVARY AND P. DESNUELLE, *Biochim. Biophys. Acta*, 21 (1956) 349.
263. P. SAVARY AND P. DESNUELLE, *Biochim. Biophys. Acta*, 31 (1959) 26.
264. F. H. MATTSON AND E. S. LUTTON, *J. Biol. Chem.*, 233 (1958) 868.
265. H. J. AST AND R. J. VANDER WAL, *J. Am. Oil Chemists' Soc.*, 38 (1961) 67.
266. M. H. COLEMAN, *J. Am. Oil Chemists' Soc.*, 38 (1961) 685.

(1) C. Desaga, GmbH, Hauptstrasse 60, Heidelberg, Germany.
(2) Research Specialities Co., 200 South Garrard Blvd., Richmond, Calif., U.S.A.
(3) Shandon Scientific Co. Ltd., 65 Pound Lane, London, N.W. 10, England.
(4) Camag A. G., Hamburger Str. 24, Muttenz, B. L., Switzerland.
(5) Camlab Ltd., Cambridge, England.
(6) Fluka, A. G. ,Buchs, S. G., Switzerland.
(7) Macherey, Nagel & Co., Düren, Rhld., W. Germany.
(8) E. Merck, A.G., Darmstadt, Germany.
(9) M. Woelm, Eschwege, W. Germany.
(10) Mallinckrodt Chemical Works, St. Louis 7, Mo., U.S.A.
(11) Aluminium Ore Coy., East St. Louis, Ill., U.S.A.
(12) Sephadex G. 25, 200–400 mesh, Deutsche Pharmacia GmbH, Frankfurt, a.M.I, Germany.
(13) Farbwerke Hoechst A.G., Frankfurt-M., Germany.
(14) Factice "31-B Coarsely Ground", Carter-Bell Manufacturing Co., Springfield, N.J., U.S.A.
(15) Driprint HC 241B (F Speed), Eugene Dietzgen, Co., 407 10th. St. N.W. Washington, D.C., U.S.A.
(16) Photovolt Corporation, 95 Madison Avenue, New York 16, N.Y., U.S.A.
(17) "Chromoscan" photodensitometer, Joyce, Loebl & Co., Gateshead, England.

STANDARDISED CHROMATOGRAPHIC DATA: A SUGGESTION

M. BRENNER, G. PATAKI AND A. NIEDERWIESER

Institut für Organische Chemie der Universität Basel (Schweiz)

SUMMARY

Owing to questionable reproducibility, R_F-data are of limited value. In many cases this drawback may be overcome in a simple manner.

On air-dried thin layers of silica* and in solvents containing hydroxylic components liquid–liquid partition is simulated. Hence, numeric values of $R_M = \log(1/R_F - 1)$ may be used to characterise R_F as a function of an apparent partition coefficient and a parameter reflecting such properties as the quality of the silica. The difference $(R_M)_i - (R_M)_{st} = (R_k)_{i/st}$ (the suffix i referring to any substance and the suffix st to a standard material run in the system at hand) is equal to $\log(K_{st}/K_i)$ and independent on silica quality. For $(R_k)_{i/st}$ the term "chromatographic number" (chromatographische Kennzahl) is proposed. It is a useful quantity for comparison or prediction of $(R_F)_i$ on all qualities of silica for which $(R_F)_{st}$ is known. A most astonishing accuracy of calculated R_F is demonstrated by experimental data. The advantage of $(R_k)_{i/st}$ over the often used $(R_F)_{rel} = (R_F)_i/(R_F)_{st}$ is obvious. Of course, R_F-measurements must be performed in a reproducible[1] manner and a possible influence of frontal analysis of mixed solvents should not be overlooked[2]. Further experience on $(R_k)_{i/st}$ could easily accumulate from the current literature if authors supplemented their published R_F-data by supplying the R_F of a generally available standard, all observations referring to identical conditions. It is also suggested to test the applicability of this simple procedure on systems less likely to simulate partition chromatography.

REFERENCES

1. M. BRENNER, A. NIEDERWIESER, G. PATAKI AND A. R. FAHMY, *Experientia*, 18 (1962) 101.
2. M. BRENNER, A. NIEDERWIESER, G. PATAKI AND R. WEBER, in E. STAHL, *Dünnschicht-Chromatographie*, Springer Verlag, Berlin, Göttingen, Heidelberg, 1962, p. 118.

---

* Relative humidity of the atmosphere must always be the same and should not exceed 60 %; a yer thickness and drying time must also be standardised.

CHROMATOGRAPHY ON THIN LAYER OF STARCH WITH REVERSED PHASES

JIŘÍ DAVÍDEK

Chemical Technological University, Praha (Czechoslovakia)

1. INTRODUCTION

The principle of reversed phases in thin-layer chromatography was first utilized by KAUFMAN *et al.*, who used silica-gel plates impregnated with undecane for the separation of higher fatty acids.

The great potential usefulness of chromatography with reversed phases in the analysis of lipophilic substances has induced us to look for a suitable carrier for fixation of the phases used for partition of various substances significant in the food industry. On the basis of orientation trials soluble starch was found the most satisfactory medium for the purpose. All sorts of phases which are currently used in partition-paper chromatography can be successfully fixed on starch. The principle was successfully used in the separation of fat-soluble food dyes, vitamins, carbonyl compounds, and especially of aliphatic aldehydes and ketones in the form of 2,4-dinitrophenyl hydrazones.

2. EXPERIMENTAL

The impregnated chromatographic plates are prepared as follows: the impregnating substances are dissolved in a suitable organic solvent at an appropriate concentration for the chromatographic purpose (which depends primarily on their properties), mixed with soluble starch, and the suspension thus obtained is applied to glass plates in usual manner. After evaporation of the solvent the plates are ready for the chromatographic process. We found that the chromatographic separation depends not only on the concentration of the impregnating substance, but also on the time and temperature of drying of the chromatographic plates. The latter conditions are important in the case of volatile substances, *e.g.* the fixation of dimethyl formamide. With this and similar substances the drying temperature and time of development of the plates must be strictly controlled, otherwise the quality of partition separation and the R_F values become irreproducible.

When fixing paraffin oil or a vegetable oil, the drying time at room temperature does not substantially influence the quality of the chromatographic separation.

For separations of fat-soluble food dyes we used starch plates with fixed (stationary) paraffin oil. The solvent mixtures currently used in the paper chromatography proved fully satisfactory. Very good results were obtained in the identification of multicomponent mixtures, using both adsorption chromatography on thin layers of alumina, and partition chromatography on starch with fixed paraffin oil. In this way it was possible to detect even the components of mixtures which hitherto could not be identified with certainty. Similar results were obtained in the application of the method to the separation of fat-soluble vitamins. In this case again paraffin oil and various vegetable oils were used as the non-polar phase.

For the separation of carbonyl compounds in the form of 2,4-dinitrophenyl hydrazones, we used starch with fixed dimethyl formamide; the method also worked well in the separation of complex mixtures of carbonyl compounds isolated from food materials. As already mentioned, it is essential to observe strictly the working conditions in the preparation of the plates, above all the time and temperature of drying, when using the technique just described. Under the given conditions the method gives very good results, incomparably better than the separation on alumina which was used in parallel experiments for the same purpose and in the same arrangement.

The advantage of starch lies above all in that the plates prepared with it in the manner described are compact and do not necessitate careful handling. They can also be stored much more easily.

We are continuing this work for the separation of other substances significant in food technology, and the results will be reported elsewhere.

CENTRIFUGAL CHROMATOGRAPHY

XII. CENTRIFUGAL THIN-LAYER CHROMATOGRAPHY*

JAN ROSMUS, MIROSLAV PAVLÍČEK AND ZDENĚK DEYL

Central Research Institute of Food Industry, Praha-Smíchov and Automation Department, Technical University, Praha-Dejvice (Czechoslovakia)

1. INTRODUCTION

Our present knowledge concerning the acceleration of chromatographic processes by centrifugal force has led to the conclusion that in the case of paper chromatography, centrifugal force is an universal means of acceleration which does not influence the quality of the separation. From the character of the process of chromatography it is obvious that there is no reason why centrifugal force, apart from accelerating the flow of the mobile phase, should influence the quality and course of separation on arbitrary supporting material. This assumption was verified by a centrifugal arrangement of thin-layer chromatography on silica gel containing binder.

The realization of centrifugal development of thin-layer chromatograms was hindered by some difficulties due to the fundamental problem of centrifugal chromatography *i.e.* the distribution of the mobile phase. From all hitherto reported distributors of the mobile phase, the central-type distributor[1] seems to be the most useful for this special case. This is mainly because eccentric feeding of the mobile phase according to McDONALD[2] results in disturbance of the adsorbent layer whilst the other available distributors cannot be applied for constructional reasons.

2. MATERIAL AND METHODS

The layer of silica gel[3] containing about 15% plaster of Paris was spread in the usual manner on aluminium discs which had a hollow in the centre of the same shape as the ball of the distributor. The lower side of the disc was equipped with a conical projection in order to fix it on the bearing head of the apparatus (Fig. 1). The development does not differ from the usual process of development of paper chromatograms in a centrifugal field; the flow of the mobile phase is, however, two to three times more rapid (depending on the thickness of the layer).

As model substances some dyes were chosen (butter yellow, sudan red and indophenol) and 2,4-dinitrophenylhydrazones of carbonyl compounds. The mobile phase

---

* For Part XI, see Z. DEYL, J. ROSMUS AND M. PAVLÍČEK, *Mikrochim. Acta*, (1963) 390.

References p. 121

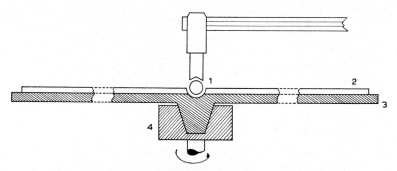

Fig. 1. Schematic view of centrifugal arrangement of thin-layer chromatography. (1) supply capillary; (2) thin layer of silica gel containing about 15% plaster of Paris; (3) aluminium disc; (4) head of chromatograph.

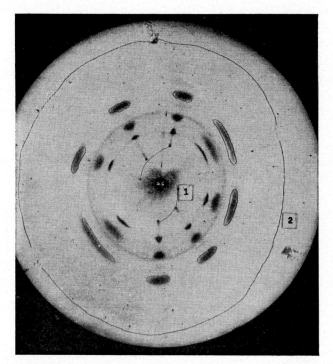

Fig. 2. Chromatographic separation of butter yellow, sudan red and indophenol. Mobile phase: chloroform. Run 5 min. The samples were spotted at four different distances from the axis of gyration. (1) start, (2) front.

used in the first case was chloroform and in the second the mixture benzene–light petroleum (1 : 1). The quality of the separations is shown in Figs. 2 and 3.

The development of a disc of 20 cm diameter took 5–10 min at 500 r.p.m.

3. SUMMARY

A technique of centrifugal thin-layer chromatography on silica gel is reported. The

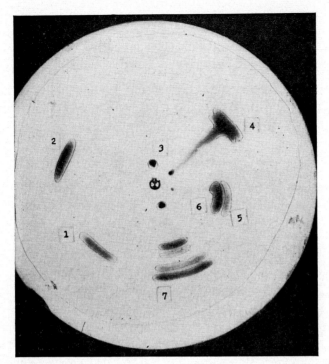

Fig. 3. Chromatographic separation of some 2,4-dinitrophenylhydrazones of carbonyl compounds. Mobile phase: benzene–light petroleum (1 : 1). Run 10 min. (1) formaldehyde, (2) acetoacetic acid ethyl ester, (3) laevulic acid, (4) fural, (5) methyl ethyl ketone, (6) acetone, (7) mixture of laevulic acid, acetone, methyl ethyl ketone, formaldehyde and acetoacetic acid ethyl ester.

process permits prompt chromatographic separation of a quality equal to that achieved by the usual technique of thin-layer chromatography.

ACKNOWLEDGEMENT

The authors wish to express their deep gratitude to Dr. L. Lábler, Institute of Organic Chemistry and Biochemistry, Czechoslovak Academy of Sciences, Prague, for the interest shown in this work.

REFERENCES

1. M. Pavlíček, J. Rosmus and Z. Deyl, *J. Chromatog.*, 7 (1962) 19.
2. H. J. McDonald, E. W. Bermes and H. G. Shepherd, *Naturwiss.*, 44 (1957) 9.
3. J. Pitra and J. Štěrba, *Chem. Listy*, 56 (1962) 545.

ÜBER DIE ANWENDUNG DER ZIRKULARTECHNIK BEIM CHROMATOGRAPHIEREN AUF KIESELGEL-DÜNNSCHICHTEN. TRENNUNG UND REINDARSTELLUNG VON MORPHIN, PAPAVERIN UND CHININ AUS DEREN GEMISCHEN

MAX VON SCHANTZ

Abt. für Pharmakognosie, Universität, Helsingfors (Finnland)

1. EINLEITUNG

In allgemeinen wird bei Trennungen auf Dünnschichten das punktförmige Auftragen der Stoffe verwendet. Es gibt aber auch andere Möglichkeiten zur Applikation der Stoffe auf die Dünnschichtplatte.

Schon IZMAILOV UND SCHRAIBER[1], die die Vorarbeit für die Dünnschichtchromatographie gemacht haben, verwenden eine Form von Zirkulartechnik zur Trennung von Stoffgemischen und auch MEINHARD UND HALL[2] benutzen eine ähnliche Technik. STAHL[3,4] hat drei verschiedene Verfahren für die Zirkulartechnik auf Dünnschichten entwickelt. Eine einfache Technik ohne Trennkammer eignet sich nach STAHL gut zur schnellen Ermittlung des geeigneten Fliessmittels. Die beiden Verfahren in geschlossenen Kammern mit oder ohne Vortrennsäule setzt der Verfasser in den Hintergrund.

In meinem Laboratorium ist jedoch ziemlich viel mit der Zirkulartechnik in geschlossener Kammer gearbeitet worden (vgl. v. SCHANTZ und Mitarb.[5]) und wir haben bei dem Verfahren viele Vorteile von der Technik des punktförmigen Auftragens gefunden.

2. HERSTELLUNG DER DÜNNSCHICHTPLATTEN

Die Herstellung der Dünnschichtplatten erfolgte mit einem besonderen, für grössere Platten passenden, aus dem gewöhnlichen Desaga-Gerät entwickelten Gerät in folgender Weise.

Mit Hilfe von Chromschwefelsäure gereinigte Glasplatten, 40 cm × 40 cm, die in der Mitte eine Durchbohrung von ø 1.5–2.0 mm hatten, wurden mit einer Kieselgelmasse bestrichen, bereitet durch Verrühren von Kieselgel G mit etwa der doppelten Wassermenge in einem Mörser. Wenn dickere Schichten erwünscht werden, muss verhältnismässig mehr Streichmasse genommen werden. Bei der Trennung von Filix-Phlorogluciden (v. SCHANTZ[7]) wurde anstatt des Wassers eine geeignete Pufferlösung (pH 6) verwendet. Als Streichgerät wurde das in Abb. 1 und 2 abgebildete Gerät

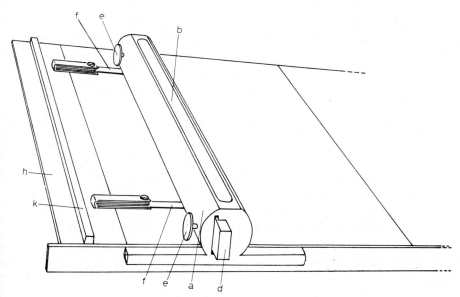

Abb. 1. Streichgerät für Dünnschichtplatten 40 cm × 40 cm auf dem Streichtisch.

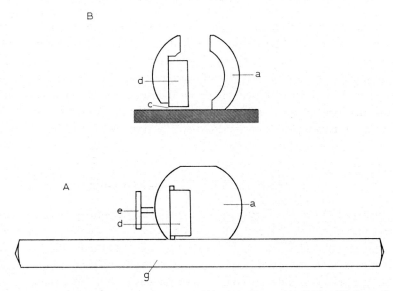

Abb. 2. Streichgerät für Dünnschichtplatten 40 cm × 40 cm. (A) Ende, (B) Querschnitt.

verwendet. Es besteht aus einem dickwandigen, schweren, 43 cm langen, an beiden Enden geschlossenen Metallrohr (a) das sowohl unten als oben eine 2 cm breite und 40 cm lange Spalte (b) für Füllung und Auslauf der Streichmasse besitzt. An der unteren Seite ist etwa 5 mm der hinteren Rohrwand beseitigt (Abb. 2B), so dass wenn das Gerät auf einer Scheibe steht, zwischen Rohr und Scheibe eine 5 mm hohe und 40 cm breite Spalte (c) frei wird. Zum Bestreichen wird das Gerät also in Richtung

Literatur S. 131

von links nach rechts geführt (Abb. 1). Innen an der Hintenwand des Rohres ist eine unten wenigstens 1 cm dicke in vertikaler Richtung bewegbare Metallscheibe (d) eingefällt, die mit Schrauben (e) festklemmbar ist. Mit Hilfe dieser Scheibe kann man die Schichtdicke einstellen. Verwendet man, wie in Abb. 1 gezeigt ist, plane Metallstücke (f), deren Dicke genau gemessen ist, kann die Schichtdicke mit Hilfe von diesen gut eingestellt werden. Das an dem einen Ende des Rohres befestigte Metallstück (g) dient als Gleitstück für das Gerät. Als Schablone wird ein Tisch, 150 × 50 cm benutzt, der an zwei grenzenden Seiten eine nach aussen mit Plast überzogene Metallliste hat. Auf den Tisch wird erst eine 40 cm lange und 10 cm breite Glasplatte (h) gelegt, an die eine 40 cm lange und 0.5 cm dicke Schaumgummiliste (k) befestigt ist, um das Ausrinnen der Masse vor dem Streichen zu verhindern. Auf dieser Schablone können drei Platten auf einmal bestrichen werden.

Die Platten werden nach dem Bestreichen in üblicher Weise in einem Gestell im Trockenschrank bei 105–120° getrocknet und können nach dem Trocknen in einem Behälter über Blaugel bis zur Anwendung, höchstens jedoch zwei Tage, aufbewahrt werden.

3. APPLIZIEREN DER SUBSTANZ AUF DIE PLATTE

Die zu untersuchenden Stoffe und Stoffgemische wurden im allgemeinen als 10%-ige Lösung an der Mitte der Platte durch vorsichtiges langsames Auftropfen appliziert und die über dem Loch liegende Kieselgelschicht wird erst dann vorsichtig entfernt. Man kann auch so verfahren, dass die Kieselgelschicht über dem Loch zuerst weggenommen wird, wonach man den Finger an die untere Seite des Loches setzt und das Loch mit dem zu untersuchenden Stoffgemisch füllt, wobei es am Rande des Loches von der Kieselgelschicht der Platte aufgesaugt wird. In dieser Weise kann man jedoch nicht quantitativ arbeiten.

4. ENTWICKELN DER CHROMATOGRAMME

17.5 cm von der Mitte der Platte wird an einem 2 cm breiten Ring das Kieselgel abgekratzt und die Platte hier mit Siliconfett bestrichen. Nach der Anweisungen von STAHL[4] wird ein Wattebausch durch das Loch gezogen und dann die Platte mit der Kieselgelschicht nach unten auf eine Schale von 35 cm Durchmesser und 10 cm Höhe gesetzt. In der Schale befindet sich das Laufmittel, das jetzt am Wattebausch entlang in die Kieselgelschicht strömt. Wenn die Strömung nicht sofort in Gang kommt kann man ein wenig Laufmittel von oben her mit einer Pipette auf den Wattebausch tröpfeln.

5. SICHTBARMACHUNG DER CHROMATOGRAMME UND IDENTIFIZIERUNG DER STOFFE

Nach beendetem Chromatographieren wurden die Platten getrocknet und mit einem geeigneten Reagenz besprüht oder am UV-Licht untersucht.

Die Identifizierung der Stoffe lässt sich mittels der Punkttechnik durch Chromatographieren der Reinstoffe neben dem zu untersuchenden Gemisch in üblicher Weise vornehmen. Hierbei muss man natürlich genau dasselbe Laufmittel wie bei der Zirkulartechnik verwenden. Hierbei ergibt sich die Reienfolge der Ringe des Zirkularchromatogramms. Liegen die Ringe jedoch sehr nahe an einander werden sie besser direkt am Zirkularchromatogramm identifiziert. Dabei wird zuerst ein Chromatogramm des zu untersuchenden Gemisches in üblicher Weise mittels Sprühreagenz sichtbar gemacht und das Mengenverhältnis jeder zu identifizierenden Komponente abgeschätzt. Dann werden von dem zu untersuchenden Gemisch ebenso viele Teile genommen wie zu identifizierende Komponenten vorliegen und zu je einem Teil wird die abgeschätzte Stoffmenge je einer der Komponenten zugegeben und in üblicher Weise chromatographiert und sichtbar gemacht. Durch Vergleich mit dem Chromatogramm des zu untersuchenden Gemisches ohne Zusatz von Reinsubstanz ergibt sich der Ring der zu identifizierenden Komponente.

6. ANWENDUNG DER ZIRKULARTECHNIK

Die Zirkulartechnik kann mit Vorteil zum Charakterisieren verschiedener für pharmazeutische Zwecke benutzte Stoffgemische verwendet werden, deren genaue Analyse grosse Schwierigkeiten bietet.

Die R_F-Werte der Stoffe sind hierbei nicht so gut erreichbar wie bei der Punkttechnik mit vertikaler Laufrichtung, da bei der Zirkulartechnik die R_F-Werte von der Laufzeit abhängig sind. Mit der letztgenannten Technik lassen sich aber die inneren Ringe ausdehnen, was besonders von Bedeutung ist beim Suchen von Begleitstoffen, die leicht von anderen Stoffen verdeckt werden.

Über die Verwendung der Methode zur Sichtbarmachung von kleineren Mengen von Stoffen in Stoffgemischen, die bei den üblichen Verfahren mittels der Punkttechnik nicht zum Vorschein kommen, wird in einer Untersuchung über Lavendelöle[8] am besten veranschaulicht.

Durch die Möglichkeit, grössere Stoffmengen auf Dünnschichtplatten durch die Zirkulartechnik zu trennen, ergibt sich auch, dass diese Technik zur präparativen Isolierung von Stoffen aus Stoffgemischen geeignet ist. Hierbei verfährt man auf folgender Weise.

Nachdem das Fliessmittel bis zu einem Radius von bestimmter Länge gelaufen ist, wird die Dünnschichtplatte von der Fliesskammer entfernt und sofort oder nach möglichst kurzer Trocknung bei Zimmertemperatur mit einer Glasscheibe bedeckt, vom dem ein Sektor so ausgeschnitten ist, dass das Loch der mit Kieselgel bedeckten Scheibe genau an Spitze des Sektors kommt. Die Platten werden mit Sprühreagenz behandelt, wobei nur der nicht bedeckte Sektor entwickelt wird. Danach wird die obere Glasscheibe entfernt und die unentwickelten Ringe werden im UV-Licht mit einer spitzen Nadel markiert. Bei Stoffen, die nicht im UV-Licht sichtbare Ringe liefern, müssen die Zonen unter Verwendung von einen Passer in demselben Abstand von der Mitte markiert werden.

Literatur S. 131

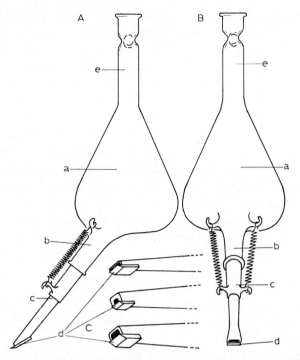

Abb. 3. Gerät zum Abkratzen der Ringe eines Zirkular-Chromatogrammes. (A) von der Seite gesehen, (B) von oben, (C) Mundstücke.

Das Abkratzen der Ringe geschieht mit Hilfe des in Abb. 3 abgebildeten Gerätes. Dieses besteht aus einem etwa 50 ml fassenden, schräg gestellten Glasbehälter, an dessen unteren Seite ein Rohr (b) durch Normalschliff mit dem austauschbaren Mundstück (c) versehen ist. Dieses besteht aus einem Messingrohr, dessen unterer Ende eine schräg gestellte Klinge (d) zum Abkratzen der Ringe von der Dünnschichtplatte trägt. Wenn das obere Rohr (e), das mit einem nicht zu dichten Wattebausch verstopft ist, mit einer Wasserstrahlpumpe verbunden wird, wird das Kieselgel des Rings in den Behälter eingesäugt und kann in dieser Weise schnell zum Auslaugen gebracht werden. Zum Abkratzen verschiedenbreiter Ringe lassen sich Mundstücke mit verschiedengrossen Öffnungen verwenden (Abb. 3C).

7. AUSLAUGEN UND KRISTALLISIEREN DER STOFFE

Das abgekratzte Kieselgel wird durch ein geeignetes Lösungsmittel ausgelaugt, das nicht nur gute Löslichkeitseigenschaften aufweisen soll, sondern auch so stark polar sein muss, dass der an das Kieselgel adsorbierte Stoff davon abtrennen wird. Die Flüssigkeit wird vom Kieselgel abzentrifugiert und das Kieselgel wenigstens fünf Mal erneut mit Lösungsmittel behandelt. Um Reste des Kieselgels aus dem vereinigten Filtrat zu entfernen wird im Vakuum auf ein kleines Volumen eingedampft und der

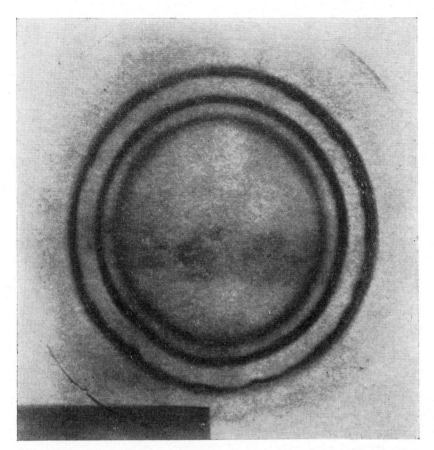

Abb. 4. Ringchromatogramm von Morphin, Papaverin und Chinin, völlig entwickelt.

Rest zwei Mal von neuem aufgelöst und zentrifugiert wie schon GÄNSHIRT UND MORIANZ[9] sowie GÄNSHIRT, KOSS UND MORIANZ[10] bei der Auswertung von Dünnschichtchromatogrammen als nötig gefunden haben.

Die so behandelten vereinigten Lösungen werden vorsichtig in einem Vakuum-Rotationsverdampfer bis zum Fällung schnell eingedampft. Durch vorsichtiges Erwärmen wird die Fällung schnell in Lösung gebracht und zur Kristallisation in den Eisschrank gesetzt. Die Kristalle werden von der Mutterlauge abzentrifugiert und vorsichtig mit wenig Lösungsmittel einige Male gewaschen.

8. TRENNUNG VON MORPHIN, PAPAVERIN UND CHININ

Als Beispiel einer Isolierung mittels der Zirkulartechnik an Kieselgel-Dünnschichten wird hier eine Reinisolierung von Morphin, Papaverin und Chinin aus deren Mischung beschrieben (Abb. 4). Als Ausgangsstoffe dienten gewöhnliche käufliche Alkaloidsalze, die nicht chromatographisch ganz rein waren.

Literatur S. 131

9. EXPERIMENTELLES

Morphin

Reinheitsprüfung. 9 mg käufliches Morphinhydrochlorid · $3H_2O$ in 5%-iger Lösung eines Gemisches von Chloroform/Äthanol 3 : 1 wurden auf eine Dünnschichtplatte, 40 cm × 40 cm, deren Schichtdicke 0.6 mm betrug (115 g Kieselgel G „Merck" + 240 ml Wasser), getropft, und das Chromatogramm mittels Methanol/Aceton/25%-ige NH_3, 100 : 45 : 5 bis zu einem Radius der Lösungsmittelfront von 14 cm chromatographiert. Das Chromatogramm wurde erst im UV-Licht betrachtet und dann mit dem von BÄUMLER UND RIPPSTEIN[11] auf Dünnschichten verwendeten Dragendorffs Reagenz* sichtbar gemacht. Der Ring des Morphins erscheint orangerot. Durch Messung des Radius zur Mitte des Rings ergab sich $R_F = 0.54$. Im UV-Licht sieht man sofort nach dem Chromatographieren einen viel kleineren leuchtenden Ring mit $R_F = 0.73$, der sich von Spalt- oder Umlagerungsprodukten des Morphins herrührt.

Wiedergewinnung nach der Adsorption an Kieselgel. 16.9 mg Morphinhydrochlorid · $3H_2O$ wurden mit 1.8 g Kieselgel G gut durchgemischt und mit 5 ml des obenerwähnten Laufmittels befeuchtet und 2 Stunden an der Luft stehen gelassen, wobei das Laufmittel verdunstete. Die Masse wurde dann quantitativ mittels 20 g eines Gemisches von Chloroform/Äthanol/25%-iges NH_3, 3 : 1 : 0.2 in ein Zentrifugenrohr gespühlt, wo die Lösung von Kieselgel abgetrennt wurde. Das letztgenannte wurde noch 5 Mal mit derselben Menge Elutionsmittel behandelt, die vereinigten Lösungen noch mit Wasser gewaschen und im Vakuum-Rotationsverdampfer bis zur Trockenheit eingeengt. Der Rückstand wurde in 0.1 N HCl gespült und die erhaltene Lösung genau auf 400 ml verdünnt und das Morphingehalt kolorimetrisch nach ADAMSON UND HANDYSIDE[12] unter Verwendung von reinem Morphinhydrochlorid · $3H_2O$ in 0.1 N HCl als Standardlösung bestimmt. Hierzu dienten:

a. Durch Elution erhaltene Morphinlösung.
b. Durch genau gleiche Behandlung mit 0.1 N HCl erhaltene Elutionslösung von Kieselgel ohne Morphin.
c. 16.3 mg reines Morphinhydrochlorid · $3H_2O$ in 400 ml 0.1 N HCl.

Je 5 ml der obigen Lösungen wurden in graduirte 25 ml-Messzylinder mit eingeschliffenen Stopfen abpipettiert, 2 ml Natriumnitritlösung zugesetzt, 15 Sek geschüttelt und genau 15 Min stehen gelassen, mit 3 ml 10%-iger NH_3-Lösung alkalisch gemacht und mit Wasser auf 12.5 ml aufgefüllt. Nach genau 5 Min Stehen wurde die Extinktion gegen die Vergleichslösung b. unter Verwendung des Blaufilters 400–450 μ bestimmt. Der Gehalt an Morphinhydrochlorid wurde in üblicher Weise durch Vergleich mit einer Verdünnungsserie der Standardlösung bestimmt und zu 86% des Ausgangsmaterials befunden.

---

* Dragendorffs Reagenz: (a) 0.85 g Wismutsubnitrat in 25%-iger Essigsäure. (b) 8 g KJ in 20 ml Wasser; 5 ml (a) und 5 ml (b) werden vermischt, 20 ml Eisessig zugefügt und mit Wasser auf 100 ml verdünnt.

Chinin

Reinheitsprüfung. 5 mg käufliches Chininhydrochlorid in 10%-iger Äthanollösung wurden auf einer Dünnschichtplatte, 40 cm × 40 cm, deren Schichtdicke 0.6 mm betrug, getropft und das Chromatogramm mittels Methanol/Aceton/25%-iges NH_3, 100 : 45 : 5 wie beim Morphin chromatographiert, im UV-Licht betrachtet und mit Dragendorffs Reagenz sichtbar gemacht. Der Ring des Chinins fluoresziert stark im UV-Licht und erscheint nach Sichtbarmachung orangerot. $R_F = 0.66$. Unmittelbar innerhalb dieses findet sich ein gleichgefärbter, viel schwächerer Ring $R_F = 0.58$, der im UV ebenfalls eine deutliche bläuliche Fluoreszenz aufweist. Dieser stammt vermutlich von Chinin- oder Chinchonin-Isomeren, womit käufliches Chinin oft beigemengt ist.

Wiedergewinnung nach der Adsorption an Kieselgel. 13.0 mg Chininhydrochlorid · $2H_2O$ wurden mit 2.5 g Kieselgel G gut durchgemischt und mit 5 ml des obengenannten Laufmittels befeuchtet, dieses an der Luft wieder verdunstet, mit Chloroform/Äthanol/25%-iges NH_3 3 : 1 : 0.2 extrahiert und gleich wie beim Morphin behandelt. In der erhaltenen, von Kieselgelresten gereinigten Lösung wurde der Chiningehalt nach PRUDHOMME[13] kolorimetrisch ermittelt. Eine Vergleichslösung wurde auch wie beim Morphin durch Elution des Kieselgels ohne Chinin bereitet.

Standardlösung: 13 mg reines Chininhydrochlorid · $2H_2O$ in 100 ml H_2O.

10 ml der obigen Lösungen wurden in einen 25 ml-fassenden Scheidetrichter mit eingeschliffenen Stopfen abpipettiert, 2 ml Puffer und 1 ml Eosinlösung* zugesetzt, drei Mal mit je 2.5 ml Chloroform kräftig geschüttelt, die Chloroformauszüge gewaschen, mit Alkohol auf 10 ml aufgefüllt und die Extinktion gegen Vergleichslösung unter Verwendung des Blaufilters 400–450 μ bestimmt. Der Gehalt an Chininhydrochlorid wurde in üblicher Weise durch Vergleich mit einer Verdünnungsserie der Standardlösung bestimmt. Ergebnis 96% des Ausgangsmaterials.

Papaverin

Reinheitsprüfung. 9 mg käufliche Papaverinbase wurden in 10%-ige Chloroformlösung wie früher beschrieben auf die Platte appliziert, mit Methanol/Aceton/25%-iges NH_3 100 : 95 : 5 auf einen Radius von 14 cm chromatographiert, im UV-Licht betrachtet und mit Dragendorffs Reagenz entwickelt. Es findet sich nur ein Ring, $R_F = 0.77$ der im UV-Licht nach einige Minuten gelbbraun fluoresziert.

Wiedergewinnung nach der Adsorption an Kieselgel. 19.9 mg Papaverinbase wurden mit 2.9 g Kieselgel G gut durchgemischt. Wie schon bei Morphin und Chinin beschrieben wurde die Masse mit dem Laufmittel befeuchtet, die Salzsäure-Lösung bereitet und auf 50 ml verdünnt. Die kolorimetrische Bestimmung wurde nach SOBOLEVA[14] von folgenden Lösungen gemacht:

---

* 0.2 g Eosin (Tribromfluorescein-Na) wurden in 70 ml H_2O gelöst und mit Chloroform in Portionen von 50 ml bis zur Farblösigkeit des Chloroformauszuges extrahiert. Die Wasserschicht wurde abgetrennt und auf 100 ml verdünnt.

Literatur S. 131

a. Durch Elution erhaltene Papaverinlösung.
b. Vergleichslösung, durch Elution von Kieselgel ohne Papaverin erhalten.
c. Standardlösung: 19.9 mg Papaverinbase in 50 ml 0.1 N HCl.

Je 2 ml der obigen Lösungen wurden in einer kleinen Glasschale auf dem Wasserbad zur Trockenheit eingedampft. Der Rückstand wurde mit 2 Tropfen 35%-iger Formaldehydlösung und 0.2 ml 80%-iger Schwefelsäure versetzt und unter öfteren Umrühren 30 Min stehen gelassen. Mit Hilfe von wenig Wasser wurde der Rückstand in einen graduierten Messzylinder übergeführt, mit 0.5 ml Bromwasser versetzt, gut durchgeschüttelt und 4 Min stehen gelassen. Dann wurden 5 ml Alkohol und 1 ml NH_3 (25%) zugegeben, auf 25 ml mit Wasser verdünnt, schnell im Eis–Salz-Gemisch durchgekühlt und nach 5 Min die Extinktion gegen die Vergleichslösung unter Verwendung des Blaufilters 400–450 μ ermittelt. Der Gehalt an Papaverinbase wurde in üblicher Weise durch Vergleich mit einer Verdünnungsserie der Standardlösung bestimmt und zu 92% des Ausgangsmaterials befunden.

Isolierung der Alkaloide aus deren Gemischen

Um die Bildung von schwer messbaren Fällungen zu vermeiden, wurde das Vermischen von Morphinhydrochlorid · $3H_2O$, Chininhydrochlorid und Papaverinbase direkt auf der Platte durch Zutropfen von folgenden Lösungen vorgenommen:

a. 10 mg Morphinhydrochlorid · $3H_2O$ in 5%-iger Lösung von Chloroform/Äthanol 3 : 1.
b. 10 mg Chininhydrochlorid in 10%-iger Äthanollösung.
c. 10 mg Papaverinbase in 10%-iger Chloroformlösung.

Das Gemisch wurde in üblicher Weise durch Verwendung von Methanol/Aceton/ 25%-iges NH_3, 100 : 45 : 5 bis zu einem Radius der Lösungsmittelfront von 14 cm chromatographiert, das Chromatogramm im UV-Licht betrachtet und die Ringe mit einer spitzen Nadel markiert. Die Kontrolle der Identität der Ringe wurde noch durch Chromatographieren eines völlig gleichen Gemisches von je 15 mg der obengenannten Alkaloide neben den Reinstoffen unter Verwendung desselben Laufmittels auf einer gewöhnlichen kleinen Platte mittels der Punkttechnik vorgenommen. Hierbei wurde die Reihenfolge Morphin, Chinin und Papaverin von innen nach aussen bestätigt. Zur weiteren Kontrolle wurde ein kleiner Sektor des Chromatogramms unter Verwendung der früher erwähnten Glasscheibe sichtbar gemacht und jeder Ring mittels des in Abb. 3 abgebildeten Gerätes abgekratzt.

Kristallisation von Morphin

(a) Das Morphin wurde wie vorher aus dem Kieselgel 5 Mal mit je 20 g eines Gemisches von Chloroform/Äthanol/25%-iges NH_3, 3 : 1 : 0.2 extrahiert und in folgender Weise zur Auskristallisation weiter verarbeitet.

Die vereinigten Auszüge von 4 Platten wurden in einer Vakuum-Rotationsverdampfer zur Trockenheit eingedampft und 0.6 ml 85%-iger Äthanol unter vorsichtigem Erwärmen bis zur Auflösung zugetropft. Die etwas trübe Lösung wurde zentrifugiert, die Mutterlauge vorsichtig mittels einer Kapillare abgesaugt, nochmals zur Trockenheit eingedampft, aufgelöst und wieder durch Zentrifugieren gereinigt. Die so erhaltene Lösung wurde nochmals zur Hälfte eingedampft und im Eisschrank zum Kristallisieren gebracht. Durch Absaugen der Mutterlauge und Waschen der Kristalle mit 50%-igem Äthanol ergab sich nach Trocknen als Ausbeute 5.9 mg Morphinbase vom Schmp. 241–242°.

(b) Das Morphin von 6 Platten wurde 5 Mal mit je 10 ml 0.1 N HCl sofort aus dem abgekratzten Kieselgel extrahiert. Der Auszug, etwa 50 ml wurde mit 3.5 ml 2 N NaOH und 0.5 g Ammoniumsulfat versetzt und 5 Mal mit 8 ml Chloroform ausgeschüttelt. Die Chloroformschicht wurde mit Wasser Gewaschen, mittels Na_2SO_4 getrocknet, im Vakuum-Rotationsverdampfer eingedampft und weiter wie oben unter (a) geschrieben behandelt. Ausbeute 9.7 mg Kristalle vom Schmp. 241–242°.

Kristallisation von Chinin

Die abgekratzten Chininringe von 6 Platten wurden 5 Mal mittels 10 ml 0.1 N H_2SO_4 extrahiert, in einen Scheidetrichter mit 2 ml 2 N NaOH Lösung unter Zusatz von 0.5 g Ammoniumsulfat gebracht und mit Chloroform sorgfältig ausgeschüttelt. Nach Waschen und Trocknen der vereinigten Chloroformphasen wurde mit Na_2SO_4 getrocknet und im Vakuum-Rotationsverdampfer zur Trockenheit eingedampft. Der Rückstand wurde in 0.09 ml Äthanol mit 10%-iger H_2SO_4 angesäuert, durch vorsichtiges Erwärmen gelöst und im Eisschrank zum Kristallisieren gebracht. Durch Absaugen der Mutterlauge und Waschen der Kristalle mittels Äthanol ergab sich nach dem Trocknen der Kristalle als Ausbeute 10.4 mg Chininsulfat vom Schmp. 110–114°.

Kristallisation von Papaverin

Das Papaverin wurde wie vorher aus dem Kieselgel 5 Mal mit je 20 g eines Gemisches von Chloroform/Äthanol/25%-iges NH_3, 3 : 1 : 0.2, extrahiert und in folgender Weise zum Auskristallisieren weiter verarbeitet: die vereinigten Auszüge von 4 Platten wurden in einem Vakuum-Rotationsverdampfer zur Trockenheit eingedampft. Der Rest wurde in 1.22 ml 95%-igem Äthanol durch vorsichtiges Erwärmen aufgelöst und die Lösung von Resten des Kieselgels wie unter Morphin beschrieben ist, gereinigt. Ausbeute 10.3 mg Papaverinbase vom Schmp. 147-147,5°.

LITERATUR

1. N. A. IZMAILOV UND M. S. SCHRAIBER, *Farmatsiya* Nr. 3, (1938) 1.
2. J. E. MEINHARD UND N. F. HALL, *Anal. Chem.*, 21 (1949) 185.
3. E. STAHL, *Dünnschicht-Chromatographie*, Berlin, Göttingen, Heidelberg, 1962.
4. E. STAHL, *Parfuem. Kosmetik*, 39 (1958) 564.
5. M. v. SCHANTZ und Mitarb., *Planta Med.*, unter Druck.
6. E. STAHL, *Pharmazie*, 11 (1956) 633; *Chemiker Ztg.*, 62 (1958) 323; *Parfuem. Kosmetik*, 39 (1958) 564; *Pharm. Rundsch.*, 1 (1959).
7. M. v. SCHANTZ, *Planta Med.*, 10 (1962) 22, 98.
8. M. v. SCHANTZ, unter Druck.
9. H. GÄNSHIRT UND K. MORIANZ, *Arch. Pharm.*, 293/65 (1960) 1066.
10. H. GÄNSHIRT, F. W. KORS UND K. MORIANZ, *Arzneimittel-Forsch.*, 10 (1960) 943.
11. J. BÄUMLER UND S. RIPPSTEIN, *Pharm. Acta Helv.*, 36 (1961) 382.
12. D. C. M. ADAMSON UND F. P. HANDYSIDE, *J. Pharm. Pharmacol.*, 19 (1946) 350.
13. R. O. PRUDHOMME, *J. Pharm. Chim.*, (9) 1 (1940) 8.
14. O. SOBOLEVA, *Aptechn. Delo*, 4 (4) (1955) 37.

ZUR DÜNNSCHICHTCHROMATOGRAPHIE MEHRKERNIGER AROMATISCHER KOHLENWASSERSTOFFE

HANS-JOACHIM PETROWITZ

Bundesanstalt für Materialprüfung, Berlin-Dahlem (Deutschland)

1. EINLEITUNG

Im Rahmen von Untersuchungen an Steinkohlenteerölen für Holzschutzzwecke haben wir uns eingehend mit der Dünnschichtchromatographie von Phenolen[1], heterocyclischen Stickstoff-Verbindungen[2,3] und mehrkernigen aromatischen Kohlenwasserstoffen beschäftigt. Über die Chromatographie der Kohlenwasserstoffe und über Zusammenhänge zwischen dem chromatographischen Verhalten und der Konstitution einiger dieser Verbindungen soll hier berichtet werden.

Schon mehrfach ist die Analyse von mehrkernigen Kohlenwasserstoffen mit Hilfe der Säulenchromatographie oder der Papierchromatographie beschrieben worden. Ohne hier ausführlich auf die Literatur einzugehen, sollen doch als wichtige Arbeiten die säulenchromatographischen Trennungen von BEERENBLUM[4], SNYDER[5], WINTERSTEIN und Mitarbeitern[6], die Heisschromatographie von VAHRMAN[7], die mit der UV-Spektroskopie gekoppelten Analysen von LINDSEY[8] und die Publikationen von FUNAKUBO und Mitarbeitern[9] genannt werden. Mit der Papierchromatographie mehrkerniger Aromaten beschäftigten sich besonders SPOTSWOOD[10], WIELAND und Mitarbeiter[11], MICHEEL und Mitarbeiter[12], GASPARIC[13] und MALÝ[14].

Mit Hilfe der Dünnschichtchromatographie wurde bisher nur eine kleine Anzahl von Aromaten getrennt. Bei der Analyse von Isolierölen für die Elektrotechnik verwenden REY und Mitarbeiter[15] Schichten aus Kieselgel G und identifizierten neben anderen Verbindungen 2,6-Dimethylnaphthalin und Phenanthren. Als Fliessmittel wurde Hexan verwendet. JANÁK[16] trennt an Silicagel mit Hexan/Benzol (1 : 1) Anthracen, 1-Methylanthracen und Fluoren von einigen heterocyclischen Verbindungen. Schliesslich verwenden WIELAND und Mitarbeiter[17] Schichten aus acetylierter Cellulose und trennen bei Zweifach-Chromatographie mit Methanol/Äther/Wasser (4 : 4 : 1) einige mehrkernige Aromaten.

2. EINFACH- UND MEHRFACH-DÜNNSCHICHTCHROMATOGRAPHIE

Die eigenen dünnschichtchromatographischen Versuche wurden mit 16 Verbindungen an Schichten aus Kieselgel G ($\sim$250 μ) und wasserfreiem Heptan als Fliessmittel bei Kammersättigung durchgeführt. Um eine möglichst gleichmässige Aktivität der

Schichten zu erhalten, wurde die 30 Minuten betragende Trockenzeit der Platten bei 105° C genau eingehalten. Ausserdem diente bei allen Chromatogrammen Azulen als Vergleichs-Substanz. Die Laufstrecke betrug 10 cm bei einer Laufzeit von ca. 20 Minuten. Das Sichtbarmachen der Flecken erfolgte durch Besprühen der Chromatogramme mit einer Lösung von Antimon(V)chlorid in Tetrachlorkohlenstoff (20 % v/v). Einige der Substanzen lassen sich auch sehr gut im UV-Licht (254 mμ) erkennen.

Es sei hier auf die Möglichkeit hingewiesen, mehrkernige aromatische Kohlenwasserstoffe durch Besprühen mit Tetracyanoäthylen nachzuweisen, wie es TARBELL und Mitarbeiter[18] sowie PEURIFOY und Mitarbeiter[19] beschreiben. Tetracyanoäthylen ist eine äusserst aktive dienophile Komponente, und als solche geht sie die Diels–Alder-Reaktion mit grosser Leichtigkeit ein. Dabei entstehen mit den Aromaten charakteristische farbige Komplexe, z.B. blaugrün mit Anthracen, dunkelrot mit Phenanthren, hellblau mit Chrysen und rotbraun mit Pyren.

Zunächst wurden die R_F-Werte der Verbindungen als Mittelwert aus 10 Einzelmessungen bestimmt, wobei als Abweichung ± 3 angegeben werden kann. (Alle im folgenden angegebenen R_F-Werte sind mit 100 multipliziert.) Es zeigte sich, dass die aus 4 Ringen bestehenden Verbindungen R_F-Werte besassen, welche zwischen 16 (Chrysen) und 23 (Pyren) lagen. Verbindungen mit 3 Ringen schliessen sich mit Werten von 24 (Phenanthren) bis 28 (Acenaphthylen) an. Kohlenwasserstoffe, welche aus 2 Ringen bestehen, haben R_F-Werte von 28 (Diphenyl) und höher, so z.B. das Naphthalin mit 36. Die Unterschiede der R_F-Werte bei einigen Verbindungen sind nicht gross genug, um die Substanzen gut voneinander zu trennen. In solchen Fällen

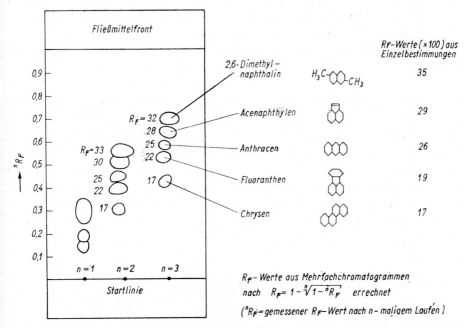

Abb. 1. Mehrfach-Dünnschichtchromatographie einiger mehrkerniger aromatischer Kohlenwasserstoffe.

Literatur S. 137

konnte mit Erfolg die Mehrfach-Dünnschichtchromatographie eingesetzt werden. Wir beobachteten gute Trenneffekte nach 2 Läufen. Ein dritter Lauf ist oft nicht notwendig, doch er verbessert die Trennung weiter, wie in Abb. 1 zu erkennen ist. Wenn die Chromatographie mit einem homogenen Fliessmittel durchgeführt wird, so gilt zum Errechnen der R_F-Werte nach n Läufen die im Bild aufgezeichnete Formel. An den Substanzflecken sind die damit errechneten R_F-Werte eingetragen. Sie zeigen eine gute Übereinstimmung mit den R_F-Werten aus Einzelbestimmungen. Neben den Trennungen, welche auf dem Bild zu sehen sind, konnten so auch die Substanz-Paare Pyren/Fluoranthen, Fluoren/Fluoranthen und Fluoren/Diphenylenoxyd getrennt werden. Allgemein lassen sich gut Zwei-Ring-Verbindungen, besonders substituierte Naphthaline, von Drei-Ring-Verbindungen trennen, was bei Einfach-Chromatographie kaum erreichbar ist. Das Diphenylenoxyd gehört zwar nicht zur Gruppe der Kohlenwasserstoffe, doch als Bestandteil des Steinkohlenteers wurde es gemeinsam mit Diphenylensulfid und Carbazol in die Untersuchung einbezogen. Bei der Einfach-Chromatographie wurde für Fluoren, Diphenylenoxyd und Diphenylensulfid übereinstimmend der R_F-Wert 24 ermittelt, während das Carbazol auf der Startlinie zurückblieb.

3. ZUSAMMENHÄNGE ZWISCHEN KONSTITUTION UND R_F-WERTEN

An einigen Beispielen sei nun der Versuch unternommen, auf die Beziehung zwischen der chemischen Konstitution und dem R_F-Wert einzugehen. Beim Betrachten der Verbindungen mit den jeweils dazugehörigen gemessenen R_F-Werten lässt sich leicht ein Zusammenhang zur Molekülgrösse erkennen. Kleine Moleküle besitzen einen hohen R_F-Wert und umgekehrt grosse Moleküle einen kleinen. Man kann versuchen, die Beziehung im Hinblick auf das Molekül in verschiedener Weise zu deuten.

WINTERSTEIN erkannte bei seinen chromatographischen Versuchen mit Diphenylpolyenen die Abhängigkeit der Adsorption von der Anzahl der konjugierten Doppelbindungen. Ähnlich kann man feststellen, dass mit zunehmender Anzahl von Doppelbindungen in mehrkernigen Aromaten bei der Dünnschichtchromatographie die R_F-Werte kleiner werden. Ein Zunehmen von Doppelbindungen im Molekül bedeutet hier aber in den meisten Fällen eine Zunahme der Anzahl der Ringe. Wenn nun die Fries'sche Regel beachtet wird, die besagt, dass die stabilste Form eines mehrkernigen Kohlenwasserstoffes diejenige ist, bei der eine grösstmögliche Zahl von Ringen die normale benzoide Anordnung von 3 Doppelbindungen besitzt, so lassen sich weitere chromatographische Effekte erklären. Als Beispiel sei auf den zwar geringen aber experimentell gesicherten Unterschied der R_F-Werte von Anthracen ($R_F = 27$) und

Phenanthren (24) hingewiesen. Beide Verbindungen bestehen aus drei Ringen und besitzen 7 Doppelbindungen. Das Phenanthren (III) besteht aus 3 benzoiden Ringen, wobei die Doppelbindung zwischen den Kohlenstoffatomen 9 und 10 olefinischen Charakter hat, wie aus dem Verhalten bei der Oxydation und besonders bei der leicht verlaufenden katalytischen Hydrierung unter Bildung von 9,10-Dihydrophenanthren geschlossen werden kann. Die so durch ihre Reaktionsfähigkeit ausgezeichnete Doppelbindung olefinischen Charakters im Phenanthren bewirkt, dass die Verbindung stärker als das Anthracen adsorbiert wird. Es wäre weiterhin denkbar, den Unterschied der R_F-Werte mit dem Vorhandensein eines chinoiden Ringes in der stabilen Form des Anthracens (I) zu erklären. Wie wir in einer anderen Arbeit feststellen konnten[1], zeigen die R_F-Werte von Hydrochinon (46) und Chinon (69) bei Verwendung des Fliessmittelsystems Benzol/Methanol/Eisessig (45:8:4) einen beachtlichen Unterschied, der nach unserer Ansicht nicht allein durch die unterschiedliche adsorbierende Wirkung der OH-Gruppen bzw. CO-Gruppen, sondern auch durch den Übergang vom benzoiden zum chinoiden System verursacht wird. Ähnlich kann es auch hier sein, zumal als weiteres Beispiel die Kohlenwasserstoffe Chrysen (16) und

Pyren (23) angeführt werden können. Beide Verbindungen bestehen aus 4 Ringen und unterscheiden sich durch die Anzahl der Doppelbindungen, die im Chrysen 9 und im Pyren 8 beträgt. Wichtiger erscheint aber für das chromatographische Verhalten zu sein, dass das Chrysen (VI) aus 4 benzoiden Ringen besteht, während im Pyren nur 3 solche vorhanden sind und die Struktur des über den nichtbenzoiden Ring verlaufenden konjugierten Systems einer 1,4-naphthochinoiden Gruppierung entspricht. Entsprechend der Fries'schen Regel muss beim Pyren die Struktur IV überwiegen. Dafür spricht auch die besondere Reaktionsfähigkeit des Moleküls in der 3- und 10-Stellung bei Substitutionsreaktionen, wobei ein Gemisch von 3,10- und 3,8-Isomeren entsteht, in welchem aber die ersteren überwiegen. Entsprechend entstehen bei der Oxydation etwa 2/3 des 3,10-Chinons und 1/3 des 3,8-Chinons. Die Anordnung der Benzolkerne in linearer oder angularer Weise (z.B. Anthracen–Phenanthren) hat auf das chromatographische Verhalten keinen besonderen Einfluss, was auch für die heterocyclischen Verbindungen Acridin (19 in Chloroform) und Phenanthridin (19) zutrifft, die sich im R_F-Wert nicht unterscheiden. (Die angulare Anordnung der Ringe im 7,8-Benzochinolin (47) führt dagegen aus sterischen Grunden zu einer beachtlichen Erhöhung des R_F-Wertes, worüber an anderer Stelle ausführlich berichtet werden soll[3].)

Literatur S. 137

Die stufenweise partielle Hydrierung einer Verbindung lässt die R_F-Werte der jeweils erhaltenen Reaktionsprodukte ansteigen. Als Beispiel dafür sei eine Verbindungs-Reihe genannt, bei der ausgehend vom Acenaphthylen (28) über das Acenaphthen (30) zum Tetrahydroacenaphthen (36) sich die R_F-Werte erhöhen. Die Aufhebung

einer Doppelbindung beim Übergang vom Acenaphthylen (VII) zum Acenaphthen (VIII) führt nur zu einer geringfügigen Verschiebung der Werte, während dagegen das Fehlen von 2 weiteren Doppelbindungen im Tetrahydroacenaphthen (IX) die Adsorptionsfähigkeit der Verbindung deutlich herabsetzt.

Wie an diesen wenigen Beispielen gezeigt werden konnte, ist es möglich, Zusammenhänge zwischen der Konstitution mehrkerniger aromatischer Kohlenwasserstoffe und ihren R_F-Werten aufzuzeigen. Im Gegensatz zu polaren Verbindungen lässt sich jedoch das chromatographische Verhalten aromatischer Kohlenwasserstoffe schwieriger deuten. Es ergab sich nun bei der graphischen Darstellung der R_F-Werte in Abhängigkeit von der Anzahl der Kohlenstoff-Atome eine einfache lineare Beziehung (Abb. 2). Die Grösse des Moleküls wird ohne Rücksicht auf die Konstitution der Verbindungen nur durch die Kohlenstoff-Anzahl ausgedrückt. Die Streuung der Punkte bei Verbindungen mit gleicher Kohlenstoff-Anzahl ist infolge der nur kleinen Unterschiede in den R_F-Werten gering. Da Methyl-Gruppen nur durch sterische

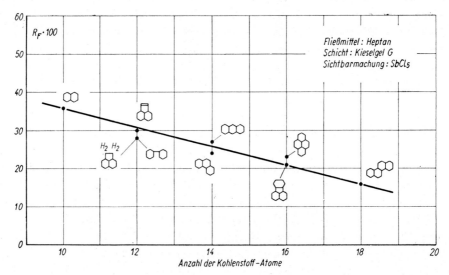

Abb. 2. Abhängigkeit der R_F-Werte von der Anzahl der Kohlenstoff-Atome bei mehrkernigen aromatischen Kohlenwasserstoffen.

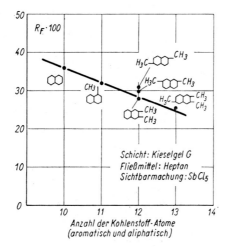

Abb. 3. Abhängigkeit der R_F-Werte von der Anzahl der Kohlenstoff-Atome bei Naphthalin und Methylnaphthalinen.

Hinderung das adsorptions-chromatographische Verhalten von Verbindungen merklich beeinflussen können[1,2], gilt die Beziehung auch für Methylnaphthaline. Wie Abb. 3 zeigt, lässt sich ebenso eine lineare Abhängigkeit des R_F-Wertes von der Summe der in den Molekülen enthaltenen aromatischen und aliphatischen Kohlenstoff-Atome zeigen. Die Einführung von Methyl-Gruppen in das Naphthalin führt zu einer Vergrösserung des Moleküls und damit zu einer Herabsetzung des R_F-Wertes. So besteht das 2,3,6-Trimethylnaphthalin (26) aus insgesamt 13 Kohlenstoff-Atomen, und der R_F-Wert liegt im Bereich der Drei-Ring-Verbindungen.

LITERATUR

1. H.-J. Petrowitz, *Erdöl Kohle*, 14 (1961) 923; G. Pastuska und H.-J. Petrowitz, *Chemiker-Ztg.*, 86 (1962) 311.
2. H.-J. Petrowitz, *Chemiker-Ztg.*, 85 (1961) 143.
3. H.-J. Petrowitz, G. Pastuska und S. Wagner, in Vorbereitung.
4. I. Beerenblum, *Nature*, 156 (1945) 601.
5. L. R. Snyder, *J. Chromatog.*, 5 (1961) 430; 6 (1961) 22.
6. A. Winterstein und K. Schön, *Z. Physiol. Chem.*, 230 (1934) 146; A. Winterstein, K. Schön und H. Vetter, *Z. Physiol. Chem.*, 230 (1934) 158; A. Winterstein und H. Vetter, *Z. Physiol. Chem.*, 230 (1934) 169.
7. M. Vahrman, *Nature*, 165 (1950) 404.
8. A. J. Lindsey, *Anal. Chim. Acta*, 20 (1959) 175; 21 (1959) 101.
9. E. Funakubo und T. Nagai, *Z. Physik. Chem.*, 220 (1962) 1; 220 (1962) 35; 221 (1962) 57.
10. T. M. Spotswood, *J. Chromatog.*, 2 (1959) 90; 3 (1960) 101.
11. T. Wieland und W. Kracht, *Angew. Chem.*, 69 (1957) 172.
12. F. Micheel und W. Schminke, *Angew. Chem.*, 69 (1957) 334.
13. J. Gasparic, *Mikrochim. Acta*, (1958) 681.
14. E. Malý, *Nature*, 181 (1958) 698.
15. E. Rey und L. Erhart, *Bull. SEV*, 52 (1961) 401.
16. J. Janák, *Nature*, 195 (1962) 696.
17. T. Wieland, G. Lüben und H. Determann, *Experientia*, 18 (1962) 420.
18. D. S. Tarbell und T. Huang, *J. Org. Chem.*, 24 (1959) 887.
19. P. V. Peurifoy, S. C. Slaymaker und M. Nager, *Anal. Chem.*, 31 (1959) 1740.

THIN-LAYER CHROMATOGRAPHY OF 2,4-DINITROPHENYLHYDRAZONES OF ALIPHATIC CARBONYL COMPOUNDS AND THEIR QUANTITATIVE DETERMINATION

GIAN MARIO NANO

Istituto di Chimica Farmaceutica e Tossicologica dell'Università, Torino (Italy)

1. INTRODUCTION

Thin-layer chromatography provides a quick and precise method for analysing some mixtures of 2,4-DNPH derivatives of aliphatic and aromatic carbonyl compounds[1,2].

TABLE 1

THIN-LAYER CHROMATOGRAPHY SEPARATIONS OF 2,4-DNPH CARBONYL COMPOUNDS

+ good separation in all solvents tested
− no complete separation
(a) separation in the eluent: nitrobenzene–chloroform–hexane (1 : 2 : 8)
(b) separation in the eluent: ether–petroleum ether b.p. 40–70° (2.5 : 7)
(c) separation in the eluent: ethyl acetate–hexane (1 : 9)
(d) separation in the eluent: nitrobenzene–carbon tetrachloride (1 : 4)
(e) separation in the eluent: nitrobenzene–cyclohexane (1 : 2)
(f) separation in the eluent: nitrobenzene–carbon tetrachloride (1 : 2)

2,4-DNPH	2,4-DNPH	$H \cdot CHO$	$CH_3 \cdot CHO$	$C_2H_5 \cdot CHO$	$C_3H_7 \cdot CHO$	$C_4H_9 \cdot CHO$	$C_5H_{11} \cdot CHO$	$C_6H_{13} \cdot CHO$	$C_7H_{15} \cdot CHO$	$HOCH_2 \cdot CHO$	$OHC \cdot CHO$	$HOOC \cdot CHO$	$(CH_3)_2CO$	$CH_2{:}CH \cdot CHO$	$(CO_2C_2H_5)_2CO$
$H \cdot CHO$			+	+	+	+	+	+	+	+	+	+	+	+	a b c
$CH_3 \cdot CHO$		+		+	+	+	+	+	+	+	+	+	+	b c d	+
$C_2H_5 \cdot CHO$		+	+		+	+	+	+	+	+	+	+	+	a	+
$C_3H_7 \cdot CHO$		+	+	+		+	+	+	+	+	+	+	+	+	+
$C_4H_9 \cdot CHO$		+	+	+	+		−	+	+	+	+	+	+	+	+
$C_5H_{11} \cdot CHO$		+	+	+	+	−		−	+	+	+	+	+	+	+
$C_6H_{13} \cdot CHO$		+	+	+	+	+	−		−	+	+	+	+	+	+
$C_7H_{15} \cdot CHO$		+	+	+	+	+	+	−		+	+	+	+	+	+
$HOCH_2 \cdot CHO$		+	+	+	+	+	+	+	+		f	−	+	+	+
$OHC \cdot CHO$		+	+	+	+	+	+	+	+	f		f	+	+	+
$HOOC \cdot CHO$		+	+	+	+	+	+	+	+	−	f		+	+	+
$(CH_3)_2CO$		+	+	+	+	+	+	+	+	+	+	+		abcde	+
$CH_2{:}CH \cdot CHO$		+	b c d	a	+	+	+	+	+	+	+	+	abcde		+
$(CO_2C_2H_5)_2CO$		a b c	+	+	+	+	+	+	+	+	+	+	+	+	

We have tried several adsorbents and several eluents and we have found that neutral alumina without binder (Woelm) is the best adsorbent for separating low-molecular weight 2,4-DNPH carbonyl compounds. The eluents containing nitrobenzene give the best separations with sharp spots[3]. The method can also be used for quantitative estimations.

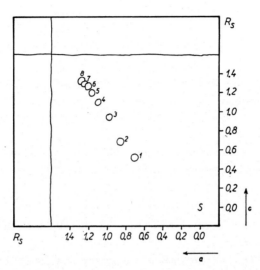

Fig. 1. Two-dimensional chromatogram of the first eight 2,4-dinitrophenylhydrazones of n-aldehydes (5 γ for every aldehyde). Solvents: (c) ethylacetate–hexane (1 : 9, v/v); (a) nitrobenzene–chloroform–hexane (1 : 2 : 8, v/v). Deep red colour by alkaline spray for all compounds: (1) formaldehyde, (2) acetaldehyde, (3) propionaldehyde, (4) butyraldehyde, (5) n-valeraldehyde, (6) caproicaldehyde, (7) oenanthol, (8) caprylaldehyde. The solvent ascended 15 cm.

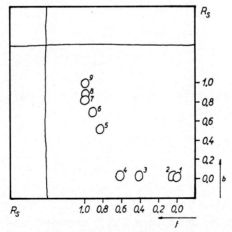

Fig. 2. Two-dimensional chromatogram of nine carbonyl 2,4-dinitrophenylhydrazine derivatives with 1, 2 and 3 carbon atoms. Solvents: (b) ether–petroleum ether 40–70° (2.5 : 7, v/v); (f) nitrobenzene–carbon tetrachloride (1 : 2, v/v). Compounds 3 and 4 give a blue colour with the alkaline spray; all other compounds give a deep red colour. (1) glycollic acid, (2) glycolaldehyde, (3) glyoxal*, (4) dihydroxyacetone*, (5) formaldehyde, (6) acetaldehyde, (7) acraldehyde, (8) propionaldehyde, (9) acetone. The compounds with an asterisk are present as osazones. The solvents ascended 20 cm.

References p. 143

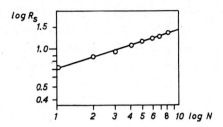

Fig. 3. Relationship between R_S and number of carbon atoms.

In Table 1 are given all the possible separations for 2,4-DNPH derivatives we have tried. Two-dimensional separations were also carried out (Figs. 1 and 2). The R_S-values, based on the running speed relative to that of 2,4-DNPH acetone are sufficient for identification.

For identifying homologous aliphatic n-aldehydes it is possible to use also a relationship between R_S and number of carbon atoms (Fig. 3):

$$R_{S_N} = B \cdot N^A,$$

where N is the number of carbon atoms of the aliphatic chain and A and B are experimental constants.

If $D_1, D_2, \ldots D_{10}$, are the distances from start for 2,4-DNPH aldehydes, we may write:

$$\log D_N = \log B + \log N \cdot A$$

and, if N is 1, respectively 10:

$$\log D_1 = \log B$$

$$\log D_{10} = \log D_1 + A$$

Thus:

$$R_{S_N} = R_{S_1} N^{\log D_{10}/D_1}$$

It is clear that two adsorption types will contribute to the movement of 2,4-DNPH n-aldehydes: the group:

$$-CH=N-NH-\underset{NO_2}{\underset{|}{\bigcirc}}-NO_2$$

is a constant contribution and it is represented by the constant B. To this, we will add the aliphatic chain contribution which, of course, varies with the number of carbon atoms.

It is also possible to make a quantitative determination of these compounds. The spots were eluted and quantitative measurements made with a spectrophotometer. We have tried this method with 2,4-DNPH formaldehyde and 2,4-DNPH acetaldehyde.

2. EXPERIMENTAL

Adsorbents

Thin layers (thickness: 300 μ) are prepared by the Desaga applicator on glass plates 20 × 20 cm. The plates were air dried and activated further by drying at 100° for 1 hour. Several adsorbents were tried as shown in Table 2.

TABLE 2

EXAMINATION OF DIFFERENT ADSORBENTS FOR SEPARATING LOW-MOLECULAR WEIGHT 2,4-DNPH CARBONYL COMPOUNDS

magnesium silicate	no separation
silica gel G	separation (incomplete)
alumina G	separation
neutral alumina*	good separation
acid alumina*	good separation**
basic alumina*	no separation

* These adsorbents are without binder (Woelm).
** Relative movement for 2,4-DNPH acetone and 2,4-DNPH propionaldehyde are reversed.

The experiments were carried out with the eluent nitrobenzene–chloroform–hexane (1:2:8, v/v) for separating mixtures of the 2,4-dinitrophenylhydrazones of formaldehyde–acetaldehyde and acetone–propionaldehyde.

Solvents

The solvents used as eluents are the commercial qualities; only nitrobenzene was redistilled with a short Widmer column.

Procedure

The elution time is one hour for eluents containing nitrobenzene. The R_S of all compounds tried in several eluents are given in Table 3. For bidimensional chromatography, it is not advisable to use nitrobenzene in the first eluent since it is strongly adsorbed by alumina and it is not easy to remove it completely from the plate.

3. PREPARATION OF THE HYDRAZONES

All 2,4-dinitrophenylhydrazones were prepared by the general procedure of O. L. BRADY[4], or by refluxing a solution of the carbonyl compound and 2,4-dinitrophenylhydrazine in ethanol without acids to prevent osazone formation from α-hydroxymonocarbonyl compounds. Generally, it is not necessary to develop the chromatoplate, but alcoholic potassium hydroxyde (10%) spray gives a deep red colour for 2,4-DNPH and a blue colour for osazones. The identification limit is 0.5–1 μg.

TABLE 3

R_S-VALUES OF DIFFERENT 2,4-DINITROPHENYLHYDRAZONES OF CARBONYL COMPOUNDS IN SEVERAL ELUENTS

(a) nitrobenzene–chloroform–hexane (1 : 2 : 8, v/v)
(b) ether–petroleum ether b.p. 40–70° (2.5 : 7)
(c) ethyl acetate–hexane (1 : 9)
(d) nitrobenzene–carbon tetrachloride (1 : 9)
(e) nitrobenzene–cyclohexane (1 : 2)
(f) nitrobenzene–carbon tetrachloride (1 : 2).
The R_S for compounds with an asterisk are for the osazones.

2,4 DNPH of	R_S-value					
	a	b	c	d	e	f
formaldehyde	0.70	0.52	0.52	0.74	0.80	0.84
acetaldehyde	0.86	0.69	0.69	0.84	0.94	0.92
propionaldehyde	0.97	0.95	0.96	0.96	0.98	1.00
butyraldehyde	1.10	1.11	1.09	1.05	–	–
n-valeraldehyde	1.16	1.19	1.21	1.08	–	–
caproicaldehyde	1.22	1.26	1.26	1.12	–	–
oenanthol	1.27	1.30	1.31	1.14	–	–
caprylaldehyde	1.30	1.34	1.34	1.17	–	–
glycolaldehyde	0.04	0	0	0.02	0.06	0.05
glyoxal*	0.10	0	0	0.17	0.67	0.42
glycollic acid	0	0	0	0	0	0
acraldehyde	0.86	0.92	0.91	0.93	–	1.01
glyceraldehyde	–	–	–	0.05	–	0.07
dihydroxyacetone*	–	–	–	–	0.72	0.65
acetone	1	1	1	1	1	1
pyruvic acid	0	0	0	0	0	0
mesoxalic acid (diethyl ester)	–	–	–	0.72	–	0.87

4. QUANTITATIVE DETERMINATION

Calibration curves were obtained using 0.5 % chloroformic solutions of the 2,4-DNPH of pure aldehydes. 1–10 μl of these solutions are placed on the neutral alumina plate

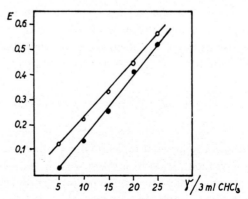

Fig. 4. Typical calibration curves (see also Table 4). (—o—o—o— 2,4-DNPH acetaldehyde, —●—●—●— 2,4-DNPH formaldehyde).

TABLE 4

STATISTICAL ANALYSIS OF DETERMINATIONS OF 10 μg OF THE 2,4-DINITROPHENYLHYDRAZONE OF ACETALDEHYDE IN PRESENCE OF AN EQUAL QUANTITY 2,4-DINITROPHENYLHYDRAZONE OF FORMALDEHYDE

$\overline{X}$ = mean.
d = average deviation = $\dfrac{\Sigma X_i}{n}$ where X_i is the deviation of individual values from the mean $\overline{X}$ and n is the number of determinations.
s = standard deviation = $\sqrt{\dfrac{\Sigma (X_i)^2}{n-1}}$.

Extinction at 355 mμ (sol. in 3 ml CHCl₃)	$\overline{X}$	d	s	$s\%$
0.200, 0.210, 0.225 0.225, 0.185, 0.220	0.211	0.013	± 0.016	8

and chromatographed with nitrobenzene–chloroform–hexane (1 : 2 : 8, v/v) as eluents. The chromatoplate was then air dried, the area of yellow spots scraped out and eluted by chloroform (generally 10 ml). Chloroformic eluates were centrifuged and the clear solutions evaporated. Finally the residue was made up to 3 ml with chloroform and the extinction was read at 343 mμ for 2,4-DNPH formaldehyde and at 355 mμ for 2,4-DNPH acetaldehyde *versus* a blank obtained by elution of pure alumina under the same conditions. This is necessary since it is not possible completely to eliminate the nitrobenzene from the adsorbent. In Fig. 4 are shown two typical calibration curves under the above conditions. In Table 4 is given the standard deviation for these determinations.

5. SUMMARY

Using thin-layer chromatography on neutral alumina, several low-molecular weight aliphatic carbonyl compounds have been separated in several solvent systems; many of these eluents contained nitrobenzene. It is possible to extend the method to quantitative determination.

REFERENCES

1. J. H. DHONT AND C. DE ROY, *Analyst*, 86 (1961) 74.
2. E. F. L. J. ANET, *J. Chromatog.*, 9 (1962) 291.
3. G. M. NANO AND P. SANCIN, *Experientia*, 19 (1963) 323.
4. O. L. BRADY, *J. Chem. Soc.*, (1931) 756.

THIN-LAYER CHROMATOGRAPHY OF STEROIDAL BASES AND HOLARRHENA ALKALOIDS

LUDVÍK LÁBLER and VÁCLAV ČERNÝ

Institute of Organic Chemistry and Biochemistry, Czechoslovak Academy of Science, Praha (Czechoslovakia)

1. INTRODUCTION

For a long time we have investigated in our laboratory the chemistry of synthetical steroidal bases and also the chemistry of steroidal alkaloids isolated from the bark of Indian shrub Holarrhena antidysenterica. As there were a relatively great number of different compounds at our disposal, we decided to study the influence of substitution of the steroidal skeleton with basic groups with special regard to stereochemistry, also from the aspect of thin-layer chromatography.

2. TECHNIQUE

Although alumina has efficient separation properties, it still does not give satisfactory results in the differentiation of some finer structural features. We therefore turned our attention to silica gel. This adsorbent, because of its acidic character can be used for chromatography of bases only under certain conditions. For such a case, either the so-called alkaline layers have to be used *i.e.* the slurry is prepared in dilute sodium hydroxide solution instead of water as has been reported by STAHL[1], or normal silica gel is used and the mobile phase contains a certain amount of strong organic base *e.g.* diethylamine according to WALDI *et al.*[2]. We obtained satisfactory results with silica gel containing 15 % of gypsum using relatively low polar mobile phases which beforehand were shaken with aqueous concentrated ammonia solution.

Chromatography was carried out in jars containing a dish with concentrated ammonia solution[3]. The separation was very satisfactory and in most cases a length of run of 10 cm was sufficient. The R_F values, as expected, varied from case to case, and therefore it was necessary to use a standard for each chromatogram. This technique, slightly modified, was used for column chromatography as well and this allowed us to isolate some minor alkaloids from Holarrhena antidysenterica in sufficient amounts to elucidate their structure[4].

3. DISCUSSION OF RESULTS

The Holarrhena alkaloids are substituted in position 3,18 and 20 of the pregnane skeleton which is either saturated or contains a 5,6-double bond. We studied the influence of substitution on these carbon atoms together with configurational relationships with respect to chromatographic behaviour.

In the first place, we investigated stereoisomeric dimethylamino groups in position 3 in $\triangle^5$, 5α and 5β-systems. For this purpose, besides Holarrhena alkaloids, we studied also some steroidal bases prepared by synthesis and derived from cholestane.

From Fig. 1 it may be seen that, according to the literature[5-8] the axial 3α-dimethylamino-$\triangle^5$-derivatives (1, 7, 9, 11) are less polar than the corresponding equatorial 3β-derivatives (4, 8, 10, 13). In the case of the saturated systems, the axial 3α-dimethylamino-5α-derivatives (2, 12) and 3β-dimethylamino-5β-derivatives (6) are less polar than the corresponding equatorial 3β-dimethylamino-5α-compounds (5, 14) and 3α-dimethylamino-5β-derivative (3).

When we turn our attention to the relationship between unsaturated and saturated compounds, we can see that in the 5α-series, the equatorial saturated 3β-derivative

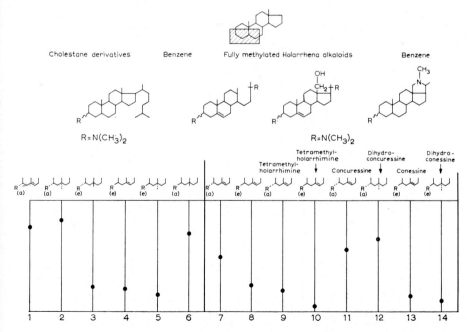

Fig. 1. TLC of 3-dimethylamino steroids. Silicagel; solvent: benzene (saturated with aqueous ammonia); detection: Dragendorff reagent; run: 10 cm; applied 5 µg.

References p. 147

(5, 14) is more polar than the parent equatorial unsaturated compound (4, 13). On the other hand, the axial 3α-dimethylamino-5α-derivative (2, 12) is less polar than the axial unsaturated compound (1, 11). The situation is the same in the 5β-series. Here again the equatorial 5β-derivative which in this case is 3α (3) is more polar than the equatorial unsaturated compound (1); accordingly, the axial 3β-derivative (6) is less polar than its parent $\triangle^5$-compound (4). When we compare all three types of compounds having the same configuration on carbon atom 3, we can see that the polarity grows in the case of 3α-derivatives from 5α < $\triangle^5$ < 5β, whereas for the 3β-derivatives the relationship is reversed i.e. 5β < $\triangle^5$ < 5α. The sequence of all the compounds is as follows: 3α, 5α < 3α, $\triangle^5$ < 3β, 5β < 3α, 5β < 3β, $\triangle^5$ < 3β, 5α.

The stereochemistry of the dimethylamino group in position 20 of the pregnane skeleton was investigated on three pairs of compounds substituted with different groups in position 3 (Fig. 2). In all cases it was found that the 20α-dimethylamino

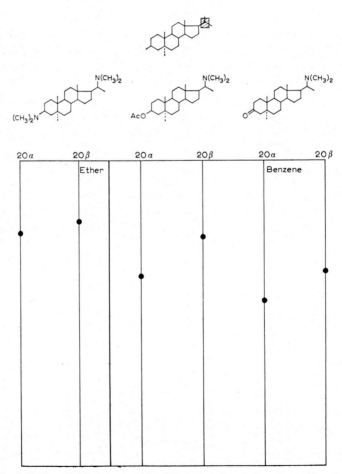

Fig. 2. TLC of 20-dimethylamino steroids. Silicagel; solvent: benzene, ether (saturated with aqueous ammonia); detection: Dragendorff reagent; run: 10 cm; applied: 5 μg.

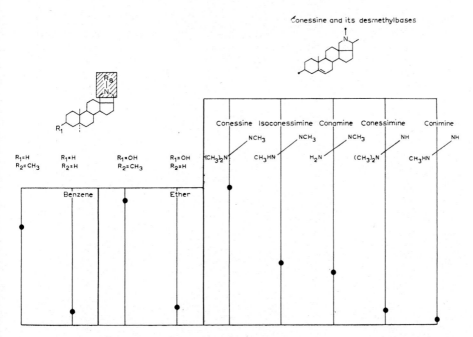

Fig. 3. TLC of conessine and its desmethylbases. Silicagel; solvent: benzene (left), ether (right) (both saturated with aqueous ammonia); detection: Dragendorff reagent; run: 10 cm (left), 17 cm (right); applied: 5 µg.

group is more polar than the 20β group. This is in agreement with the data known for the corresponding 20-hydroxy derivatives.

Further, we were interested in the degree of influence of methylation of amino groups in Holarrhena alkaloids. For this purpose, we chose the main alkaloid of Holarrhena bark — the fully methylated conessine — and its desmethyl bases (Fig. 3). From the five possible desmethyl bases four are known: isoconessimine, conamine, conessimine and conimine. As concerns the remaining fully demethylated alkaloid conarrhimine, the parent alkaloid of conessine type, its presence in the bark has not yet been convincingly proved. On the left two chromatograms we can see on synthetic model compounds the remarkably high difference between the methylated and non-methylated pyrrolidine ring closed between carbon atoms 18 and 20. This contribution is far higher than the contribution of N-methyl groups in position 3 which is only second-rate as may be seen on the right chromatogram. Here even the most polar 3-amino group in conamine which contains a methylpyrrolidine ring is still less polar than conessimine containing a relatively low polar dimethylamino group but non-methylated pyrrolidine ring.

REFERENCES

1. E. STAHL, *Arch. Pharm.*, 292/64 (1959) 411.
2. D. WALDI, K. SCHNACKERZ AND F. MUNTER, *J. Chromatog.*, 6 (1961) 61.
3. L. LÁBLER AND V. ČERNÝ, *Collection Czech. Chem. Commun.*, 28 (1963) 2932.

4. L. Lábler and F. Šorm, *Collection Czech. Chem. Commun.*, 28 (1963) 2345.
5. R. Tschesche and P. Otto, *Chem. Ber.*, 95 (1962) 1144.
6. V. Černý, J. Joska and L. Lábler, *Collection Czech. Chem. Commun.*, 26 (1961) 1658.
7. S. Heřmánek, V. Schwarz and Z. Čekan, *Collection Czech. Chem. Commun.*, 26 (1961) 1669.
8. R. Neher, *Chromatographie von Sterinen, Steroiden und verwandten Verbindungen*, Elsevier Publishing Co., Amsterdam, 1958.

THIN-LAYER CHROMATOGRAPHY OF ALKALOIDS ON MAGNESIA CHROMATOPLATES

ENRICO RAGAZZI, GIOVANNI VERONESE AND CARLA GIACOBAZZI

Institute of Pharmaceutical Chemistry, University of Padua (Italy)

1. INTRODUCTION

Today, thin-layer chromatography employs a wide range of adsorbents, according to the substances to be resolved and the degree of resolution required. As well as the classical adsorbent for thin-layer chromatography, silica gel (usually with addition of gypsum as binder), one may advantageously use alumina, kieselguhr, cellulose powder and, for particular uses, polyamide, substituted celluloses, Sephadex and many others.

Magnesium oxide has been used mixed with Celite for the column chromatography of carotenoids[1-12] and has been tried by itself, in thin layers only, by KIRCHNER *et al.*[13], who tested the possibility of separating the components of essential oils. They obtained "soft" layers, which were not suitable for the required resolution.

In the present paper magnesium oxide has been tried by itself in thin layers for separating several alkaloids. Although not extensively used, it may be regarded as a suitable adsorbent to replace alumina in many instances, even though its characteristics vary, sometimes greatly, with the process of preparation[14].

On account of its basic character, this adsorbent appears to be suitable for chromatography of basic or of neutral substances. We have therefore tried to accomplish a thin-layer separation of alkaloids which had already been studied on thin layers of silica gel and, to a minor extent, on alumina[15]. Attempts were also made to achieve quantitative determination of the separated alkaloids.

Among the various kinds of adsorbent tested, hydrated magnesium oxide, known as "magnesium oxide heavy hydrated", appears to be the most satisfactory. Although the separations obtained were satisfactory, the shape of the spots (tendency to expansion and to tailing), did not appear comparable with those obtainable with other adsorbents such as silica gel or alumina.

The impregnation of the layer with salts was then studied. As it is well known, this technique improves the separation of some classes of substances on adsorbents such as silica gel or kieselguhr. Among the salts tested, good results were obtained with magnesium sulfate and calcium chloride. The spots were compact, round, and without tailing.

The most suitable thickness of the adsorbent layer appeared to be between 0.2 and 0.3 mm.

References p. 154

Many solvent systems were studied; the results obtained with some of these are shown in Table 1.

TABLE 1

R_F VALUES OF SOME ALKALOIDS ON THIN LAYERS OF MAGNESIA

Solvent systems:
S_1 = n-hexane–acetone (3 : 1)
S_2 = n-hexane–acetone (3 : 2)
S_3 = n-hexane–acetone (4 : 1)
S_4 = n-hexane–acetone (9 : 1)
S_5 = n-hexane–ethyl acetate (9 : 1)
S_6 = ethyl acetate
S_7 = ethyl acetate–acetone (4 : 1)
S_8 = ethyl acetate–acetone–chloroform (2 : 2 : 1)
S_9 = ethyl acetate–pyridine (10 : 1)
S_{10} = chloroform–ethyl acetate (3 : 1)
S_{11} = chloroform–ethyl acetate (1 : 1)
S_{12} = chloroform–ethyl acetate (1 : 3)

Alkaloids	R_F values in											
	S_1	S_2	S_3	S_4	S_5	S_6	S_7	S_8	S_9	S_{10}	S_{11}	S_{12}
magnesia impregnated with calcium chloride												
morphine	0.03								0.05	0.05	0.14	0.03
codeine	0.16								0.29	0.44	0.64	0.32
ethylmorphine	0.30										0.74	
papaverine	0.60								1	1	0.85	0.98
thebaine	0.38								0.62	0.82	0.85	0.57
narcotine	1								1	1	0.92	1
narceine	0								0	0	0	0
quinine							0.54	0.71				
quinidine							0.60	0.80				
cinchonine							0.42	0.68				
cinchonidine							0.47	0.59				
colchicine		0.35										
aconitine				0.18			0.18	0.83				
cocaine						0.22						
hydrastine				0.63	0.31							
magnesia impregnated with magnesium sulfate												
morphine						0.35				0.28	0.20	
codeine						0.80				0.85	0.80	
papaverine						0.90				0.96	1	
thebaine						1				1	1	
narcotine						1				1	1	
narceine						0				0	0	
strychnine	0.58											0.88
brucine	0.24											0.70

The amount of each alkaloid which can be chromatographed varied between 50 and 60 µg; for certain alkaloids, such as codeine, papaverine and thebaine as much as 100 µg had to be used. The smallest quantity, which coincides with the smallest amount of alkaloid detectable on chromatoplates by the reagents (see later) was generally 3–5 µg.

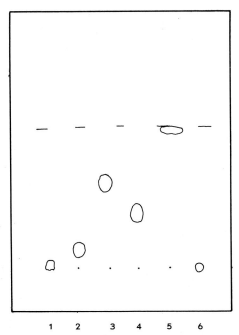

Fig. 1. Thin-layer chromatogram of six opium alkaloids in the solvent system n-hexane–acetone (3 : 1). (1) morphine; (2) codeine; (3) papaverine; (4) thebaine; (5) narcotine; (6) narceine.

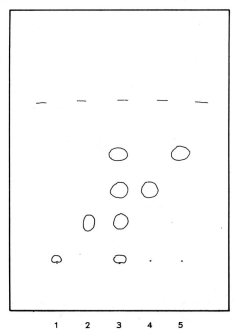

Fig. 2. Thin-layer chromatogram of some opium alkaloids; solvent system: n-hexane–acetone (3 : 1). (1) morphine; (2) codeine; (3) mixture of 1, 2, 4 and 5; (4) ethylmorphine; (5) papaverine.

References p. 154

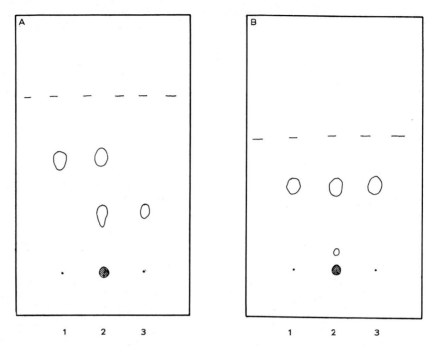

Fig. 3. Thin-layer chromatograms of drug extracts. (A) (1) strychnine; (2) Nux Vomica liquid extract; (3) brucine. Solvent system: n-hexane–acetone (3 : 1). (B) (1), (3) hydrastine; (2) Hydrastis liquid extract. Solvent system: n-hexane–acetone (4 : 1).

Figs. 1, 2 and 3 show some of the chromatograms obtained.

The chromatoplates prepared using the type of magnesia mentioned above showed high adherence of the adsorbent to the glass plate and excellent stability without binders.

The use of such thin layers of magnesia permitted a chromatographic separation either by applying alkaloids to the plate as free base dissolved in an organic solvent, or as salts in aqueous solution. Owing to the basic character of the adsorbent, the alkaloid salts were turned to free base *"in situ"* and as such were then resolved by the solvent. It was also possible to achieve a satisfactory separation of alkaloids present in extracts of vegetable drugs by applying extracts directly to the plate without any preliminary treatment. This was possible because of the behaviour (mentioned above) of the alkaloid salts, while interfering components of the extracts, especially the coloured materials, were strongly adsorbed at the point of application.

Thin layers of magnesia also permit quantitative evaluations to be achieved. The chromatographed alkaloid may be eluted with a suitable solvent from the adsorbent removed from the glass plate by means of the special extractor[16]. The zone in which the spot is located may also be dissolved in dilute hydrochloric acid, giving a clear solution. In both cases the alkaloid may be estimated spectrophotometrically. The determination of alkaloids after chromatography on thin layers of magnesia will be the subject of another note.

Finally, we wish to emphasise that this adsorbent should be properly standardised. The various kinds of magnesium oxide available on the market behave differently when used for chromatographic work: the differences can be quite large. Further work on this problem is proceeding.

2. EXPERIMENTAL

Materials

Adsorbent: magnesium oxide heavy hydrated (C. Erba n. 200301).

Solutions for impregnation of the layer: 2.5% calcium chloride ($CaCl_2 \cdot 6H_2O$) or 2% magnesium sulfate ($MgSO_4 \cdot 7H_2O$) aqueous solution.

Alkaloids: the alkaloids studied in this work are listed in Table 1.

An "Agla" micrometer syringe was used for the application of the samples to chromatoplates.

Preparation of plates and development

For a plate of 15 × 20 cm, 5 g of magnesia suspended in 9 ml of one of the above mentioned solutions was used. (For the choice of the salt solution, see Table 1.) The layer was then allowed to dry in air, horizontally. After 24 hours the chromatoplates were ready for use.

If the chromatoplates were not used at once they were kept in a desiccator or an other airtight container since the layers of adsorbent are sensitive to the carbon dioxide in air and to acid vapours.

Application of the samples to the plates and development (ascending technique) were performed according to the usual procedure for thin-layer chromatography. Development lasted two to three hours for a distance of 12–15 cm, depending on the solvent system used.

Detection

The following reagents were found to be suitable for the detection of the alkaloids studied:

(1) cobalt thiocyanate[17] obtained by dissolving 3 g of ammonium thiocyanate and 1 g of cobaltous nitrate in 20 ml of water. It produced blue spots on a white or pale pink background with these alkaloids. The colour of the spots appeared more or less strong according to the particular alkaloid, however the spots were always easily detectable. The background turned blue, particularly when the layers were impregnated with magnesium sulfate.

(2) $N/10$ iodine produced yellow or brown spots on a yellow-brown background. It was found especially suitable for the detection of sparteine, aconitine, cocaine and cinchona alkaloids.

(3) potassium chloroplatinate[18] gave violet spots on a yellowish background with strychnine, brucine, cocaine and cinchona alkaloids, whereas with the other alkaloids it gave unstable or weak spots.

(4) the classical Dragendorff reagent gave unstable brown spots on a yellow background.

References p. 154

3. SUMMARY

Thin layers of magnesia (impregnated with calcium chloride or magnesium sulfate) were used for the chromatography of several alkaloids. The amount which can be chromatographed for each alkaloid varied generally between 5 and 50–60 μg. Owing to the basic character of the adsorbent, alkaloids may also be applied on the plate as salts which become the free base *"in situ"*. It was also possible to achieve the direct chromatography of vegetable drug extracts.

The method permits quantitative determinations.

REFERENCES

1. H. H. STRAIN, *Science*, 79 (1934) 325.
2. H. H. STRAIN, *J. Biol. Chem.*, 105 (1934) 523.
3. H. H. STRAIN, *J. Biol. Chem.*, 111 (1935) 85.
4. H. H. STRAIN, *Nature*, 137 (1936) 946.
5. H. H. STRAIN, *J. Biol. Chem.*, 127 (1939) 191.
6. H. H. STRAIN, *Ind. Eng. Chem., Anal. Ed.*, 18 (1946) 605.
7. H. H. STRAIN, *J. Am. Chem. Soc.*, 70 (1948) 588.
8. G. MACKINNEY, S. ARONOFF AND B. T. BORNSTEIN, *Ind. Eng. Chem., Anal. Ed.*, 14 (1942) 391.
9. J. V. PORTER AND F. P. ZSCHEILE, *Arch. Biochem.*, 10 (1946) 537.
10. R. E. KAY AND B. PHINNEY, *Plant Physiol.*, 31 (1956) 226.
11. M. J. SABACKY, L. B. JONES, H. D. FRAME AND H. H. STRAIN, *Anal. Chem.*, 34 (1962) 306.
12. *Official Methods of Analysis of the Association of Official Agricultural Chemists*, Washington, 1955, pp. 235 and 816.
13. J. G. KIRCHNER, J. M. MILLER AND G. J. KELLER, *Anal. Chem.*, 23 (1951) 420.
14. P. CHOVIN, in *"Chromatographie en Chimie Organique et Biologique"*, Masson et Cie, Editeurs, Paris, 1959, vol. I, p. 46.
15. D. WALDI, in *"Dünnschicht-Chromatographie"*, Springer Verlag, Berlin, Göttingen, Heidelberg, 1962, p. 287–300.
16. J. S. MATTHEWS, A. L. PEREDA V. AND A. AGUILERA P., *J. Chromatog.*, 9 (1962) 331.
17. J. DELTOMBE AND G. LEBOUTTE, *J. Pharm. Belg.*, 13 (1958) 41.
18. *Anfärbenreagenzien für Dünnschicht- und Papierchromatographie*, E. Merck AG, Darmstadt, p. 27, n. 113.

SOME APPLICATIONS OF THIN-LAYER CHROMATOGRAPHY FOR THE SEPARATION OF ALKALOIDS

G. GRANDOLINI*, C. GALEFFI, E. MONTALVO**, C. G. CASINOVI
AND G. B. MARINI-BETTÒLO

Istituto Superiore di Sanità, Department of Biological Chemistry, Rome (Italy)

1. INTRODUCTION

In a recent paper WALDI[1] reviewed the applications of thin-layer chromatography in the separation of alkaloids and suggested a systematic method of their analysis by this technique[2].

We should like to report here some further results on the separation of alkaloids which we have obtained by means of thin-layer chromatography.

2. SEPARATION OF STEREOISOMERS

In the course of our work on the alkaloids of the cyclopentanopiperidine group[3] we have found thin-layer chromatography to be a very suitable technique to solve some of the rather complicated problems connected with the separation of the natural alkaloids of *Skytanthus acutus*.

Whereas by means of paper chromatography no resolution was observed, by means of thin-layer chromatography, using silica gel and chloroform–methanol (70:30) as solvent, we demonstrated[4] that the volatile alkaloid fraction of *Skytanthus acutus* consists mainly of three saturated and one unsaturated stereoisomeric cyclopentanopiperidines.

(1) α-skytanthine   (2) δ-skytanthine   (4) β-skytanthine

The spots were detected both with Dragendorff reagent, followed by ceric sulphate in 1% sulphuric acid, and by exposure to iodine vapour.

In Table 1 the R_F values of skytanthine and the other bases are reported.

---

* Istituto di Chimica Farmaceutica, University of Perugia.
** Ecuadorian fellow at the Istituto Superiore di Sanità.

TABLE 1

R_F VALUES OF CYCLOPENTANOPIPERIDINES

β-Skytanthine	0.38
α-Skytanthine	0.58
δ-Skytanthine	0.65
Dehydroskythantine	0.76

These bases were identified by direct chromatographic comparison with the samples prepared according to a stereospecific synthesis[4].

This example of separation of stereoisomers by thin-layer chromatography demonstrates once more the great possibilities of this method.

Furthermore the chromatographic behaviour of these alkaloids on thin layers has enabled us to establish the conditions for preparative separation in silica-gel columns.

Here less polar solvents were used for elution in order to achieve better resolution; *i.e.* chloroform with 1% methanol is used followed by chloroform containing increasing percentages of methanol.

3. SEPARATION OF STRYCHNOS ALKALOIDS

The extracts of several Strychnos plants was checked by thin-layer chromatography using silica gel and butanol–HCl (95:5, equilibrated with water) or chloroform–methanol (80:20) as solvent.

An artificial mixture of several tertiary Strychnos alkaloids was used as control.

TABLE 2

R_F OF TERTIARY ALKALOIDS ON SILICA GEL

Substance	Solvent	
	butanol–HCl (+H$_2$O) (95 : 5)	CHCl$_3$–CH$_3$OH (8 : 2)
Brucine	0.37	0.46
Dehydrobrucine	0.37	0.55 (*)
Diaboline	0.47	0.23 (*)
Strychnine	0.48	0.58 (*)
Holstine	0.51	0.14 (*)
Retuline	0.58 (*)	0.20 (*)
Vomicine	0.44 (*)	0.94
N-Oxystrychnine	0.52 (*)	0.26 (*)
α-Colubrine	0.45 (*)	0.52
β-Colubrine	0.48	0.60
Ethoxystrychnine	0.60	0.95
Pseudostrychnine	0.55	0.95

(*) Average value.

The spots were detected with Dragendorff reagent and with iodine vapours.

Specific chromatic reagents were used according to MATHIS AND DUQUÉNOIS[5] in order to confirm the identification of the substances.

R_F values of the tertiary alkaloids in the two solvents are reported in Table 2.

Fig. 1 shows a separation of a natural alkaloid mixture in an arrow poison from Malaya in comparison with strychnine, diaboline and holstine using one-dimensional chromatography.

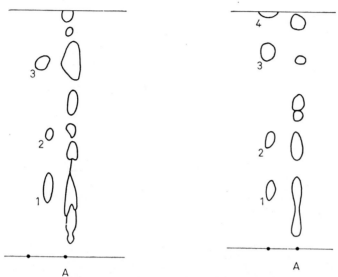

Fig. 1. One-dimensional chromatogram on silica gel. Solvents CHCl$_3$–CH$_3$OH (80 : 20). 1: Strychnine; 2: Diaboline; 3: Holstine as reference. A: Alkaloids from dart poison.

Fig. 2. A: Alkaloids from Strychnos B. 779. Solvents CHCl$_3$–CH$_3$OH (80 : 20). 1: Holstine; 2: Diaboline; 3: Strychnine; 4: Vomicine + Ethoxystrychnine as a reference.

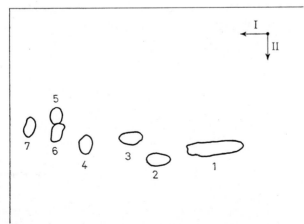

Fig. 3. Two-dimensional chromatogram of an artificial mixture of tertiary Strychnos alkaloids Silica gel; Solvent I: CHCl$_3$–CH$_3$OH (80 : 20); Solvent II: n-Butanol–HCl (95 : 5) sat. water. 1: Holstine + Retuline; 2: N-Oxystrychnine; 3: Diaboline; 4: Brucine; 5: Dihydrobrucine; 6: Strychnine; 7: Vomicine.

References p. 159

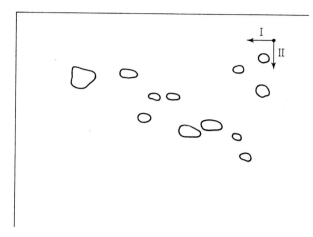

Fig. 4. Two-dimensional chromatogram of Strychnos axillaris tertiary alkaloids. Silica gel; Solvent I: $CHCl_3$–CH_3OH (80 : 20); Solvent II: n-Butanol–HCl (95 : 5) sat. water.

In Fig. 2 the separation of the alkaloids from a Strychnos species is shown, employing the same conditions.

A two-dimensional technique was also developed in order to obtain better separations, using in the first direction chloroform–methanol (80 : 20) and in the second, butanol–HCl (95 : 5, equilibrated with water).

This method permits good separations both of artificial and natural alkaloid mixtures.

Figs. 3 and 4 show some examples of this separation.

It was possible to detect in the extract of a Malayan Strychnos the presence of pseudostrychnine ethyl ether, which has not hitherto been found in plants.

It was also demonstrated that this alkaloid is probably an artifact (pseudostrychnine in contact with ethanol yields the ethyl ether spontaneously) by the use of carefully purified solvents checked by IR spectrography.

4. SCREENING OF PLANT MATERIAL AND PARTICULAR BASIC FRACTIONS FROM PLANTS

Thin-layer chromatography has proved to be of great importance in combination with paper electrophoresis for the screening of extracts of plant material.

We were able by this means to detect alkaloids in a number of botanical samples of barks from Amazonia which are used by Indians as hallucinogens (supplied to us by Prof. E. BIOCCA*).

The presence of harmine and of minor quantities of other alkaloids was demonstrated in a sample of bark (Bi.389), the alkaloid being detected both by Dragendorff reagent and U.V. light.

---

* Italian Amazonian Expedition, 1962–1963.

The identification of harmine was confirmed by direct comparison with a sample of the pure alkaloid.

Thin-layer chromatography was also used for the rapid screening of a number of samples of Strychnos extracts. This proved useful not only for the systematic survey of plant material, but also in order to get information for further preparative separation of the alkaloids.

5. SUMMARY

A number of applications of alkaloid separation by thin-layer chromatography are reported, *viz.*, separation of stereoisomers of Skytanthus alkaloids and the separation of tertiary alkaloids. This method can also be employed for screening the basic fractions of plant material.

REFERENCES

1. D. WALDI, *Arch. Pharm.*, 292 (1959) 206.
2. D. WALDI, K. SCHNACKERZ AND F. MUNTER, *J. Chromatog.*, 6 (1961) 61.
3. G. B. MARINI-BETTÒLO, C. G. CASINOVI AND F. DELLE MONACHE, *Sci. Rept. Ist. Super. Sanita*, 2 (1962) 195.
4. C. G. CASINOVI, F. DELLE MONACHE, G. B. MARINI-BETTÒLO, E. BIANCHI AND J. GARBARINO, *Gazz. Chim. Ital.*, 92 (1962) 479.
5. C. MATHIS AND P. DUQUÉNOIS, *Ann. Pharm. Franç.*, 21 (1963) 47.

THIN-LAYER CHROMATOGRAPHY OF ISOMERIC OXIMES. II

I. PEJKOVIĆ–TADIĆ, M. HRANISAVLJEVIĆ–JAKOVLJEVIĆ AND S. NEŠIĆ

Faculty of Sciences, University of Belgrade (Yugoslavia)

1. INTRODUCTION

We recently showed[1] that the isomeric α- and β-benzaldoximes, benzoin oximes and anisoin oximes could be differentiated successfully by thin-layer chromatography. After these very promising results we proceeded to the study of the isomeric oximes with the following pairs:

p-tolualdoximes,
p-anisaldoximes,
p-cuminaldoximes and
o, m, p-nitrobenzaldoximes.

From the work of other investigators it is known that the determinations of oxime configurations usually depend on the different products of Beckmann rearrangements. Since one form of some oximes changes into the other either simply by crystallization or by standing in a solution, it was obvious that this problem needed more detailed examination by a physical method where the "principle of minimum structural change"[2] could be retained. Special mention should be made of papers by PALM AND WERBIN[3] who tried to determine the configuration of isomeric oximes by IR spectra, REKKER AND VEENLAND[4] by UV spectra, LUSTIG[5] by NMR spectra and TYUTYULKOV et al.[6] by numerous polarographic measurements.

We thought it could well be useful to introduce thin-layer chromatography as another physical method which could be helpful as a correlative study for oxime configurations.

By thin-layer chromatography of the oximes mentioned we obtained very good separations of isomers on standard thin layers (according to Stahl) in a short time — up to 45 minutes; this is important because it minimizes the possible interconversion of isomers. It is also worth mentioning that the very different R_F-values were always registered in one single solvent mixture for all isomers (benzene : ethylacetate, 50 : 10, v/v).

2. RESULTS

The results obtained in the present investigations showed that all α-isomers of the derivatives we examined of benzaldehyde (*p*-tolualdoximes, *p*-anisaldoximes, *p*-cu-

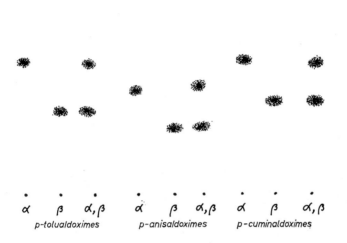

Fig. 1. Thin-layer chromatogram of isomeric oximes. Conditions: 10 μl of 0.5% (w/v) tetrahydrofuran solution of each oxime applied to the plate, absorbent: silica gel G (Merck), solvent system: benzene–ethyl acetate (50 : 10, v/v), solvent front 14 cm, spraying reagent 0.5% cupric chloride solution.

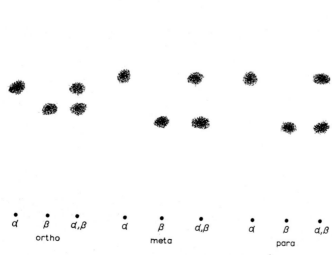

Fig. 2. Thin-layer chromatogram of isomeric nitrobenzaldoximes. Conditions are the same as those cited for Fig. 1.

References p. 164

minaldoximes and o, m, p-nitrobenzaldoximes) had higher R_F-values than the β-isomers (Figs. 1 and 2). On the other hand, with the previously examined condensation derivatives of benzaldehyde (i.e. "oin oximes", benzoin oximes and anisoin oximes) all α-isomers displayed smaller R_F-values than the β-isomers. This was not the only regularity observed: in the case of α- and β- o,m,p-nitrobenzaldoximes the R_F-values of all α-isomers showed a slight and steady increase in the order $o < m < p$-substituents, whereas the R_F-values of β-isomers showed a slight decrease in the same sense (Fig. 2 and Table 1).

TABLE 1
R_F-VALUES OF ISOMERIC OXIMES

Oximes	$R_F \cdot 100$	
	α	β
p-Tolualdoximes	54	33
p-Anisaldoximes	42	27
p-Cuminaldoximes	54	37
o-Nitrobenzaldoximes	47	40
m-Nitrobenzaldoximes	52	35
p-Nitrobenzaldoximes	53	34

We may add that the use of this type of chromatography revealed the presence of minor contaminants (ca. 0.1 γ), and in many cases great difficulty was encountered in purifying an isomer i.e. in obtaining the isomer in a single spot — the chromatographically pure isomer. Some isomers, although having a constant melting point, were not a sample of the pure isomer but a mixture of both. On the contrary, in the case of β-o-nitrobenzaldoxime the sample with m.p. 122° showed one spot and gave the same R_F-value as did the sample with the melting points given[7] in the literature 136° and 144°*. Thus, the melting point, as REKKER AND VEENLAND (loc. cit.) observed, also loses its value as a criterion of purity in our experiments.

The present investigation is expected to furnish further evidence for the examination of the "Isomerie Konstante" given by BRENNER AND PATAKI[8], and to be helpful in detecting both isomeric forms of an oxime in cases where only one form could be detected by a chemical method.

We hope to come back to the problems in question in future papers.

3. EXPERIMENTAL

Materials

The α- and β-forms of p-tolualdoxime, p-cuminaldoxime (4-isopropylbenzaldoxime) and p-anisaldoxime (4-methoxybenzaldoxime) were obtained by the method given for α- and β-benzaldoximes[9].

---

* Probably different polymorphic forms.

α-p-Tolualdoxime was twice recrystallized from the mixture benzene–petroleum ether. M.p. 77–78°.

β-p-Tolualdoxime. Two crystallizations from benzene yielded the pure isomer of m.p. 112°.

α-p-Cuminaldoxime. The crude product of m.p. 54° gave only one spot as did the product recrystallized from ether, m.p. 57°.

β-p-Cuminaldoxime. An attempted oximation according to BECKMANN[10] was unsuccessful. The product was obtained following the directions given for *β*-benzaldoxime. The crude product was twice recrystallized from ether–petroleum ether. M.p. 108–110°.

α-p-Anisaldoxime. Recrystallized twice from benzene, the compound had m.p. 61–62°.

β-p-Anisaldoxime. Recrystallization twice from benzene gave m.p. 130°.

α-o-Nitrobenzaldoxime was prepared following the procedure for *α*-furoinoxime[11]. Recrystallization from benzene–petroleum ether gave a product of m.p. 100–101°.

β-o-Nitrobenzaldoxime was synthetized according to GOLDSCHMIDT AND VAN RIETSCHOTEN[7]. After numerous extractions with warm benzene the residual material had m.p. 138° and 144°.

α-m-Nitrobenzaldoxime was prepared following the instructions for *α*-furoinoxime[11]. Recrystallization from ether–petroleum ether gave a product of m.p. 122°.

β-m-Nitrobenzaldoxime was prepared according to the procedure given for *β*-benzaldoxime[9]. The crude product was washed several times with cold benzene resulting in the pure isomer of m.p. 85–87°.

α-p-Nitrobenzaldoxime was obtained according to the procedure for *α*-furoinoxime[11]. Recrystallized twice from water the compound melted at 129–130°.

β-p-Nitrobenzaldoxime was prepared following the procedure for *β*-benzaldoxime[9], the only difference being that the *α*-isomer was converted into the *β*-form by gaseous HCl in absolute chloroformic solution instead of in ethereal solution. The crude product was dissolved in acetone, precipitated with water and immediately filtered. M.p. 180°.

Solvent system
Benzene : ethyl acetate, 50 : 10, (v/v).

Spraying reagent
Aqueous 0.5 % cupric chloride solution.

Method
The glass plates (20 × 13 × 0.5 cm) were coated with a 0.2 mm-thick layer of standardized silica gel G (according to Stahl). The plates were activated at 110° for

10 min and kept in a moisture-free chamber. The oximes were dissolved in tetrahydrofuran and 10 μl of 0.5% (w/v) solution of each oxime was applied to the plate. The spots were dropped in a line 2.5 cm from the lower edge of the plate at intervals of 2 cm. The rectangular tanks (24 × 16 × 8 cm) were saturated for 1 h and lined with filter paper. The solvent front was 14 cm, and the running time was 40–45 min.

It should be mentioned that the β-oximes, *i.e.* the oxime–metal complexes, appeared immediately as green spots whereas the metal-complexes of the α-isomers showed up as pale greenish-brown spots after heating the chromatoplate at 110° for 10 min; the spots thus obtained subsequently turned green.

4. SUMMARY

The thin-layer chromatographic method has been used for the separation and detection of α- and β-*p*-tolualdoximes, α- and β-*p*-anisaldoximes, α- and β-*p*-cuminaldoximes, α- and β-*o,m,p*-nitrobenzaldoximes. All α-isomers displayed markedly greater R_F-values than the β-isomers. In addition to this, in the case of *o,m,p*-nitrobenzaldoximes the R_F-values of all α-isomers showed slight and steady increase in the series of $o < m < p$-substituents, whereas the R_F-values of the β-isomers showed a slight decrease in the same sense.

REFERENCES

1. M. Hranisavljević–Jakovljević, I. Pejkovic–Tadić and A. Stojiljković, *J. Chromatog.*, 12 (1963) 70.
2. E. Eliel, *Stereochemistry of Carbon Compounds*, McGraw-Hill, New York, 1962, p. 95.
3. A. Palm and H. Werbin, *Can. J. Chem.*, 32 (1954) 858; 31 (1953) 1004.
4. R. F. Rekker and J. U. Veenland, *Rec. Trav. Chim.*, 78 (1959) 739.
5. E. Lustig, *J. Phys. Chem.*, 65 (1961) 491.
6. V. N. Tyutyulkov *et al.*, *Compt. Rend. Acad. Bulgare Sci.*, 14 (1959) 739; *C.A.*, 55 (1961) 25549f.
7. H. Goldschmidt and W. H. van Rietschoten, *Ber.*, 26 (1893) 2101.
8. M. Brenner and G. Pataki, *Helv. Chim. Acta*, 44 (1960) 1420.
9. A. Vogel, *A Textbook of Practical Organic Chemistry*, Longmans, Green and Co, London, 1948, p. 683.
10. E. Beckmann, *Ber.*, 37 (1904) 3044.
11. A. Werner and Th. Detscheff, *Ber.*, 38 (1905) 78,69.

THE ADAPTATION OF THE TECHNIQUE OF THIN-LAYER CHROMATOGRAPHY TO AMINOACIDURIA INVESTIGATION

J. OPIEŃSKA-BLAUTH, H. KRACZKOWSKI AND H. BRZUSZKIEWICZ

The Department of Biochemistry, The Medical School and The Agricultural College, Lublin (Poland)

1. INTRODUCTION

The methods used in tests for aminoaciduria, both chromatographic and electrophoretic, are on the whole troublesome and not free from error.

The results of experiments as given by various authors using different methods do not always agree with one another. The advantages of thin-layer chromatography have also been found useful in the separation of amino acids. The first attempt to separate amino acids by the TLC-technique was made by BRENNER, NIEDERWIESER AND PATAKI[1,2]. These authors used plates covered with silica gel G; they obtained their best results in the separation using as solvent butanol–acetic acid–water (60 : 20 : 20) and in the second dimension phenol–water (75 : 25).

The above authors gave the R_F values for nearly 40 amino acids and their derivatives in 6 different solvents. The TLC-technique for the separation of amino acids in moulds was also used by MUTSCHLER AND ROCHELMAYER[3] and NÜRNBERG[4] in protein hydrolysates.

Good results in the separation of a mixture of amino acids were also obtained by WOLLENWEBER[5] using plates covered with cellulose (Merck MN 300)*.

Our tests were carried out on:

(a) 34 pure amino acids
(b) amino acid mixtures (0.01 M solutions)
(c) urine
(d) urine to which various amino acids were added.

We used the original Desaga** apparatus, plates 20 × 20 and 17 × 17 cm were covered with silica gel G according to Stahl. We carried out attempts to separate the amino acids simultaneously on plates and on Whatman No. 3 paper.

---

* We should like to thank the firm of Merck for kindly supplying us with samples of silica gel and cellulose.
** We extend our most grateful thanks to the firm of C. Desaga for the gift of an apparatus for thin-layer chromatography.

References p. 173

2. SOLVENTS

We used the following solvents to develop the chromatograms:
(1) butanol–acetic acid–water (3 : 1 : 1)
(2) phenol–water (1 : 1)
(3) propanol–water (1 : 1)

and certain special solvents for the separation of the first group of amino acids such as:

I. (a) butanol–acetic acid–water (1 : 1 : 5)
   (b) phenol–m-cresol–borate buffer of pH 9.3 (1 : 1 : 1) to separate the leucine and valine groups.

II. (a) methylethyl ketone–pyridine–acetic acid–water (70 : 15 : 15 : 2)
    (b) phenol–water (3 : 1), to separate amino acids of the leucine and valine groups.

III. To separate the basic amino acids:
    (a) acetone–pyridine–n-butanol–water–diethylamine (15 : 9 : 15 : 8 : 10)
    (b) propanol–water (7 : 3).

IV. (a) *tert.*-butanol–acetic acid–water (15 : 3 : 2)
    (b) amyl alcohol–2,4-lutidine–water (178 : 178 : 114).

To detect the chromatograms by the spray technique we used:
(1) Ninhydrin 0.2% in 95 ml acetone + 5 ml 2,4,6-collidine
(2) isatine 0.2% solution in acetone
(3) p-dimethylaminobenzaldehyde 1 g of preparation in 10 ml conc. HCl.
Before using add 4 parts of acetone to one of the mixture.

3. RESULTS

Remarks on the method

1. The development of chromatograms on plates at an angle or held vertically causes the gel to slide off. Whichever way the plates were held, the R_F values were the same. Because of this, we carried out the development of the chromatograms in glass cells with several (4–6) plates held vertically in glass racks at least 1.5–2 cm apart.

2. When we used layers of silica gel G of varying thicknesses, the determined R_F values were not uniform.

3. In comparative tests upon the original Desaga plates and upon plates of Polish glass, the important role played by the even surface of the plate in the spreading of a layer of gel of the same thickness was demonstrated by us.

4. As in paper chromatography it is essential in TLC to preserve identical conditions during development (temperature of chamber, saturation, time of development).

5. As we observed in paper chromatography cold development gives equally good

results in the plate method, for it enables the identification of some amino acids to be made on the basis of differences in staining, in depth of colour, and time of their appearance. After preliminary identification in the cold, the plates were heated to 60° in order to detect the less sensitive stains of amino acids.

6. The sensitivity of ninhydrin to various amino acids on plates with a sorbent did not differ from its sensitivity on paper. There are no visible differences in the chromatograms of mixtures of amino acids on paper and on plates. Those amino acids which do not separate on paper do not separate on glass plates (Table 1, Fig. 1).

TABLE 1

R_F-VALUES OF AMINO ACIDS IN DIFFERENT SOLVENTS (TLC)

solvents:
A: n-propanol–water (1 : 1)
B: phenol–water (3 : 1)
C: n-butanol–acetic acid–water (4 : 1 : 1)

No.	Amino acids		R_F values in solvents		
			A	B	C
1	Alanine	Ala	0.49	0.25	0.32
2	β-Alanine	β-Ala	0.49	0.20	0.33
3	Aspartic acid	Asp	0.56	0.05	0.26
4	Asparagine	Asp-NH$_2$	0.46	0.19	0.22
5	α-aminobutyric acid	α-NH$_2$Bu	0.54	0.25	0.32
6	β-aminobutyric acid	β-NH$_2$Bu	0.48	0.32	0.36
7	γ-aminobutyric acid	γ-NH$_2$Bu	0.38	0.30	0.39
8	α-aminoisobutyric acid	α-AIB	0.57	0.26	0.35
9	β-aminoisobutyric acid	β-AIB	0.46	0.29	0.38
10	Arginine	Arg	0.06	0.14	0.13
11	Citrulline	Cit	0.46	0.29	0.26
12	Cystine	(Cys)$_2$	0.08	0.09	0.07
13	Cystic acid	CySO$_3$H	0.61	0.55	0.20
14	Glutamic acid	Glu	0.55	0.07	0.32
15	Glutamine	Glu-NH$_2$	0.55	0.28	0.24
16	Glycine	Gly	0.50	0.18	0.28
17	Histidine	His	0.33	0.24	0.10
18	Hydroxyproline	Hypro	0.63	0.33	0.26
19	Isoleucine	Ileu	0.60	0.36	0.47
20	Leucine	Leu	0.63	0.37	0.53
21	Lysine	Lys	0.05	0.08	0.10
22	Methionine	Met	0.62	0.36	0.43
23	Norleucine	NLeu	0.66	0.54	0.55
24	Norvaline	NVal	0.67	0.56	0.54
25	Ornithine	Orn	0.06	0.05	0.08
26	Phenylalanine	Phe	0.63	0.41	0.54
27	Proline	Pro	0.48	0.45	0.24
28	Serine	Ser	0.52	0.19	0.29
29	Taurine	Tau	0.59	0.22	0.29
30	Threonine	Thr	0.66	0.18	0.28
31	Tryptophan	Try	0.69	0.45	0.56
32	Tyrosine	Tyr	0.65	0.36	0.50
33	Valine	Val	0.56	0.29	0.38

The separation of amino acids in urine

In preparing urine for analysis desalting is essential; this is also necessary for the

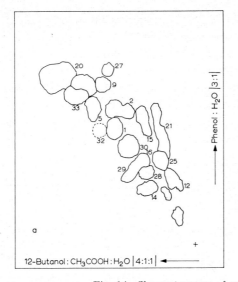 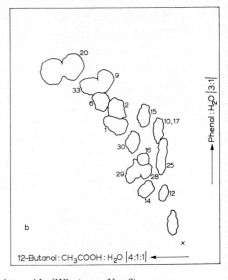

Fig. 1A. Chromatograms of amino acids (Whatman No. 3).

Numbering of amino acids in figures corresponds with numbering in Table 1.

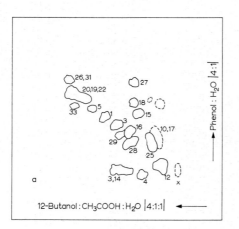 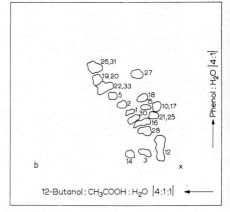

Fig. 1B. Chromatograms of amino acids (T.L.C.).

TLC method when applying more than 10 μl of urine. For desalting we used Amberlite IR 120. The removal of urea with urease appeared unnecessary. When a quantity of 1–10 μl of physiological urine is applied, no spots of amino acids can be seen on the paper but on the plates with silica gel we identified 3–6 spots with ninhydrin (Figs. 2, 3 and 4).

When the same quantities of desalted urine are applied to the paper and the plates many more spots appear on the plates (Table 2).

The best TLC chromatograms are obtained using, for non-desalted urine, not more than 10 μl of urine and for desalted urine not more than 50 μl.

TABLE 2

Quantity of desalted urine µl	Number of spots	
	Whatman No. 3	TLC
25	0	4
50	5–6	10
100	9–10	14–16*

* Spots not sharply separated.

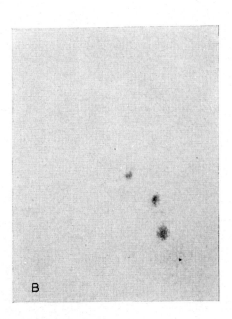
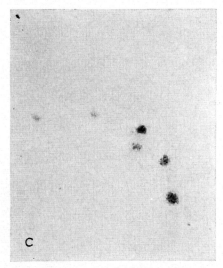

Fig. 2. Chromatograms of amino acids in urine without desalting (TLC). A: 3 µl, B: 5 µl, C: 10 µl.

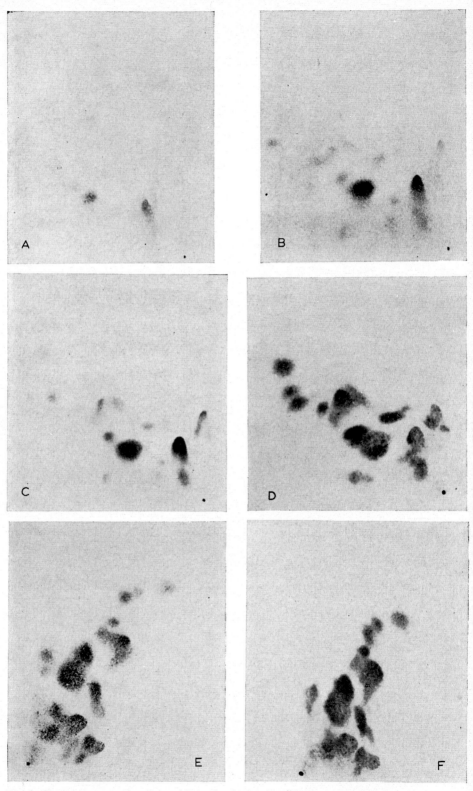

Fig. 3. Chromatograms of amino acids in desalted urine (TLC). A: 5 µl, B: 12,5 µl, C: 25 µl, D: 37.5 µl, E: 67 µl, F: 75 µl.

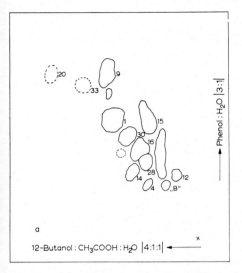
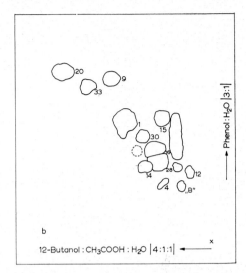

Fig. 4A. Chromatograms of amino acids in desalted urine I (a) and in desalted urine II (b) (Whatman No. 3); 100 μl.

Numbering of amino acids in figures corresponds with numbering in Table 1.

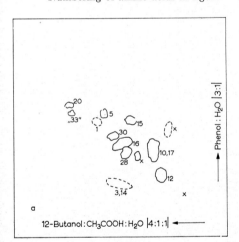
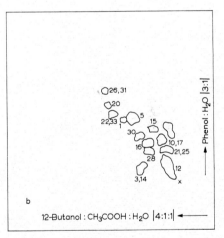

Fig. 4B. Chromatograms of amino acids in desalted urine I (a) and in desalted urine II (b) (TLC); 25 μl.

Like BRENNER AND NIEDERWIESER we observed that the solvent butanol–acetic acid–water is better than the mixture propanol–water, which is very good for separation on paper (Fig. 5).

The separation of basic amino acids (Orn, Lys, His, Arg, Cit).

When using our universal solvent with butanol and phenol we obtain only 3 spots: Orn + Lys; His + Arg; Cit. When using the propanol solvent a better separation can be obtained. Four spots were obtained: Cit; His; Arg + Lys or Lys + Orn.

References p. 173

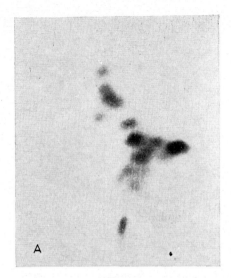

Fig. 5A. Chromatogram of amino acids in desalted urine (TLC) in solvents: n-propanol–water (1 : 1) and phenol–water (3 : 1).

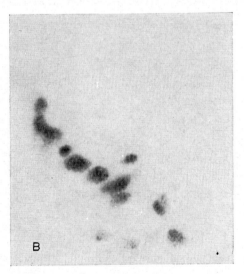

Fig. 5B. Chromatogram of amino acids in desalted urine (TLC) in solvents: n-butanol–acetic acid–water (4 : 1 : 1) and phenol–water (3 : 1).

Attempts to use a perfect separating solvent mixture for all those amino acids on paper consisting of acetone, pyridine, acetic acid and EDTA did not give good results for TLC. The violet colouring of background with ninhydrin in the presence of traces of diethylamine cannot be removed as is possible on paper with citric acid.

The identification of tryptophan

Our former attempts to detect tryptophan on paper showed that development in phenol caused considerable loss of this amino acid. Our experiments with TLC have shown that plates covered with cellulose are better for the detection of tryptophan than plates with silica gel.

The identification of aminobutyric acids (β-aminoisobutyric acid and α- and γ-aminobutyric acid)

To identify these aminobutyric acids the best solvent was propanol–water (1 : 1) in the first dimension and phenol–water (7 : 3) in the second dimension.

We obtained 3 spots. Separation by electrophoresis also gives good results.

The identification of proline and hydroxyproline

The identification was carried out with isatine, or with the aid of ninhydrin and isatine. Both these amino acids behave in the same way on paper and silica gel.

The identification of neutral and acid amino acids

The identification of these amino acids such as glycine, serine, asparagine, alanine, tyrosine, threonine, glutamine, cysteine, glutamic acid, aspartic acid does not present any difficulties in TLC.

4. SUMMARY

The TLC-method is suitable for tests for aminoaciduria and takes much less time than paper chromatography. The development of two-dimensional chromatograms can be done in 3–4 h, whereas on paper the development in the same solvents takes about 40 h. In addition, the TLC method needs a smaller volume of material than the paper technique.

Where quantities less than 10 μl are used, desalting is unnecessary, and the number of spots of amino acids detected is the same as in 50 μl of desalted urine on paper.

The amino acid spots on plates in comparison with those on paper cover a smaller area and are more sharply divided. In the separation of the various groups of amino acids giving collective stains on the paper we were unable to obtain better results in TLC with the mixtures we used.

Preliminary attempts to employ electrophoresis in pyridine buffer pH 3.9 together with chromatography gave encouraging results; especially good results can be obtained in the separation of aminobutyric acids and basic amino acids.

Attempts at quantitative determinations with glass plate chromatograms are in progress.

REFERENCES

1. A. R. FAHMY, A. NIEDERWIESER, G. PATAKI AND M. BRENNER, *Helv. Chim. Acta*, 44 (1961) 2022.
2. M. BRENNER AND A. NIEDERWIESER, *Experientia*, 16 (1960) 378.
3. E. MUTSCHLER AND H. ROCHELMAYER, *Arch. Pharm.*, 292 (1959) 449.
4. E. NÜRNBERG, *Arch. Pharm.*, 292 (1959) 610.
5. P. WOLLENWEBER, *J. Chromatog.*, 9 (1962) 369.

DIRECT ANALYSIS OF PHOSPHOLIPIDS OF MITOCHONDRIA AND TISSUE SECTIONS BY THIN-LAYER CHROMATOGRAPHY

S. B. CURRI, C. R. ROSSI AND L. SARTORELLI

Istituti di Chimica Biologica e di Anatomia Patologica della Università di Padova ed Impresa di Enzimologia del C.N.R. (Italy)

1. INTRODUCTION

All current chromatographic procedures for lipid analysis on tissues or subcellular particles involve a preliminary extraction of the lipid components with proper organic solvents. Results presented in this paper show that it is possible to resolve the lipid fractions of mitochondria or tissue sections into their components by means of chromatography on a thin layer of silica gel[1], without preliminary extraction.

2. METHODS

Preparation and chromatography of mitochondria

Rat liver mitochondria were prepared in 0.25 M sucrose. Nuclei and cell debris were sedimented by centrifugation for 8 min at 700 g. Mitochondria were sedimented from the supernatant by centrifugation for 15 min at 9000 g and washed once with 0.25 M sucrose.

A suspension of 10–20 μl of mitochondria (containing 20–40 μg of N protein) was deposited on glass paper (Whatman). Sucrose spread into a large spot while the mitochondria remained in a central small yellow spot. This spot was cut out and inserted into a silica gel plate (18 × 20 cm) in which it was firmly fixed by means of a drop of the same solvent used for the chromatographic run: chloroform/methanol/water; 70/30/5; v/v. The silica gel G "Merck" (200 μ thick) was previously activated for 1 h at 120° C.

Preparation and chromatography of tissue sections

Tissue sections (10 μ thick and 3–5 mm in diameter) were obtained by cryostatic or frozen microtome procedures. The sections were applied on cover slides (20 × 20 mm) and these fixed on chromatographic plates by means of gum arabic or Apathy's syrup. The plates were then coated with a suspension of silica gel G (200 μ thick) by means of a thin-layer applicator, allowed to stand at room temperature for 30 min and dried in a desiccator. For the chromatographic run chloroform/methanol/water (70/30/5; v/v) was used.

After the chromatographic run the cover slides were removed from the plates, washed with water and submitted to histological analysis.

Quantitative analysis of phospholipids

The spots of separated phospholipids, revealed by exposure to iodine vapour, were introduced into Pyrex test tubes. 0.4 ml of 70% perchloric acid was added to each tube and completely evaporated by gently agitating over a gas flame. This treatment led to a complete digestion of phospholipids and at the same time rendered the kieselgel completely insoluble. 2.4 ml of 12% perchloric acid was then added to the test tubes and they were placed in boiling water for 10 min, the kieselgel sedimented by centrifugation and the supernatant analyzed for phosphorus content according to WAGNER[1].

3. RESULTS

The direct chromatography of mitochondria, with exclusion of any extraction step, resulted in a clear separation of the following phospholipids (Fig. 1): phosphatidic acid, phosphatidylethanolamine + phosphatidylserine, phosphatidylcholine, phosphatidylinositol + sphingomyelin, lysophosphatidylcholine. This separation is as good

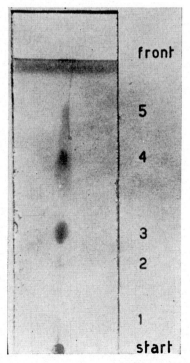

Fig. 1. Direct chromatography of rat liver mitochondria on silica gel G. (1) lysophosphatidylcholine, (2) sphingomyelin + inositolphosphatide, (3) phosphatidylcholine, (4) phosphatidylethanolamine + phosphatidylserine, (5) phosphatidic acid.

References p. 179

as that obtained with the same chromatographic method after preliminary extraction of phospholipids with methanol–chloroform[2].

In order to check the accuracy of this simplified procedure for quantitative analysis as well, the results for mitochondrial phosphatidylcholine and phosphatidylethanolamine plus phosphatidylserine obtained by the present method were compared with those obtained for methanol–chloroform extracts[2] by chromatography on silica gel, and by paper chromatography[3]. As shown in Table 1 the results obtained by all the procedures are identical.

TABLE 1

QUANTITATIVE ANALYSIS OF PHOSPHOLIPID FRACTIONS FROM RAT LIVER MITOCHONDRIA
(in μg of phospholipid P/mg protein N)
(Average of 4 experiments)

Fraction	Paper chromatography (Dawson)	Thin-layer chromatography of methanol–chloroform extracts	Direct chromatography of mitochondria on thin layers
Phosphatidylcholine	18.7	18.2	18.9
Phosphatidylethanolamine + phosphatidylserine	14.8	15.2	15.9

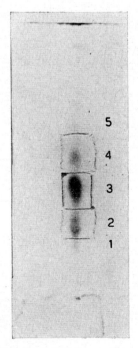

Fig. 2. Direct chromatography of rat liver sections on silica gel G. (1) lysophosphatidylcholine, (2) sphingomyelin + inositolphosphatide, (3) phosphatidylcholine, (4) phosphatidylethanolamine + phosphatidylserine, (5) phosphatidic acid.

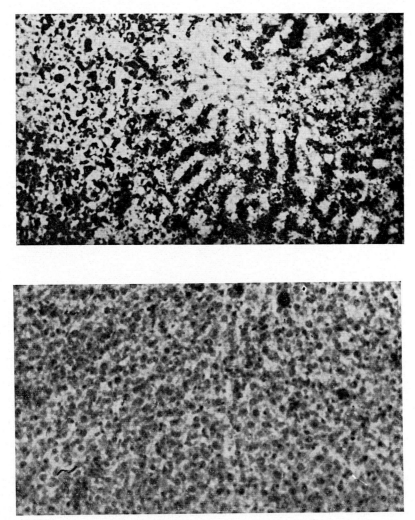

Fig. 3. Histochemical analysis of rat fatty liver sections. (A) Tissue section before chromatography, stained by Chiffelle's method for lipids. (B) Serial section of the same tissue after the chromatographic run, stained by Chiffelle's method and contrasted with nuclear staining (haematoxylin).

Direct chromatography of rat liver sections gave a clear separation of the same phospholipids found in liver mitochondria (Fig. 2).

The histochemical analysis of "fatty" liver sections (obtained from livers of rats fed on a low protein–high fat diet[4] for 20–30 days) showed, after chromatography, the complete disappearance of the lipids detectable by CHIFFELLE's method[5] as shown in Fig. 3.

Fig. 4 shows a typical pattern of phospholipids obtained by direct chromatography of a sciatic nerve.

References p. 179

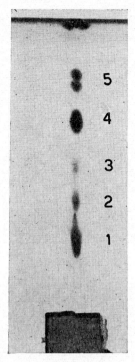

Fig. 4. Direct chromatography of a sciatic nerve section on silica gel G. (1) sphingomyelin, (2) phosphatidylcholine, (3) unidentified, (4) phosphatidylethanolamine + phosphatidylserine, (5) unidentified.

4. DISCUSSION

The direct thin-layer chromatography of mitochondria and tissue sections, with exclusion of any extraction step, resolves the lipid fractions into their components as well as does thin-layer chromatography of their organic solvent extracts.

As was shown by the correspondence of the results with those obtained by different methods, direct thin-layer chromatography can be used not only for qualitative but also for quantitative analysis. Compared with the other procedures direct thin-layer chromatography is easier, much less expensive, and is useful for the rapid screening of a large number of samples. The time required for quantitative analysis has been considerably reduced both by the elimination of the extraction step with organic solvents and by simplification of the determination of phospholipid phosphorus. By using WAGNER's method[1] both the complete digestion of phospholipids and rendering the kieselgel insoluble is achieved by heating at 180° C for 4 h in the presence of 0.4 ml of 70% perchloric acid saturated with ammonium molybdate. The same result can be obtained more rapidly by complete evaporation (in 5–10 min) of the 0.4 ml of 70% perchloric acid which is employed for each spot of the various phospholipids.

For the tissue sections, direct chromatography permits the exact location of the chromatographed lipids when the tissue sections are submitted to a proper histological analysis after chromatography. The direct chromatographic procedure, unlike the method based on a preliminary chemical extraction of lipids, does not destroy the structure of the tissues but only releases the lipid fractions which are under investigation.

5. SUMMARY

The lipid fractions of rat liver mitochondria and tissue sections (liver and sciatic nerve) have been separated by direct chromatography of these materials on a thin layer plate of silica gel G avoiding the usual extraction step. The method permits the rapid quantitative analysis of phospholipids of mitochondria and of tissue sections to be carried out.

REFERENCES

1. H. WAGNER, *Fette, Seifen, Anstrichmittel*, 63 (1961) 1119.
2. J. FOLCH-PI AND M. LEES, *J. Biol. Chem.*, 191 (1951) 807.
3. R. M. C. DAWSON, *Biochem. J.*, 57 (1954) 237.
4. C. R. ROSSI, C. S. ROSSI, L. SARTORELLI, D. SILIPRANDI AND N. SILIPRANDI, *Arch. Biochem. Biophys.*, 99 (1962) 214.
5. T. L. CHIFFELLE AND F. A. PUTT, *Stain Technol.*, 26 (1951) 51.

QUALITATIVE AND QUANTITATIVE ANALYSIS OF NATURAL AND SYNTHETIC CORTICOSTEROIDS BY THIN-LAYER CHROMATOGRAPHY

G. CAVINA AND C. VICARI

Istituto Superiore di Sanità, Department of Biology, Rome (Italy)

1. INTRODUCTION

In recent years thin-layer chromatography has been applied to an ever increasing number of analytical problems, in many cases obtaining results of great interest. (For an authoritative review see STAHL[1]).

In the field of steroid analysis we would like to mention the work of BARBIER et al.[2], NEHER AND WETTSTEIN[3], who have separated mixtures of low polar steroids of different classes; STARKA AND MALIKOVA[4] who have separated pregnanetriol and pregnanediol in urinary extracts; LISBOA AND DICZFALUSY[5] and STRUCK[6], who have studied the fractionation of estrogens and their metabolites and derivatives, and also the experimental contribution of WALDI to STAHL's volume[1].

The behaviour of some of the principal corticosteroids has been described in the solvent systems reported in Ref. 4 and by CÉRNY[7] and HERMANEK[8], while ADAMEC et al.[9] have described the behaviour of natural corticosteroids on silica gel plates without binder. LISBOA[10] has described the application of various solvent systems to the analysis of numerous C_{21}-Δ^4-3-CO-steroids.

Recently BENNET AND HEFTMANN[11] have recorded the behaviour of various natural corticosteroids in thin-layer chromatography, and MATTHEWS, PEREDA AND AGUILERA[12] have described the advantageous results obtained by thin-layer chromatography in the quantitative analysis of some steroids.

From the above literature, we have noted that thin-layer chromatography has not been applied so far to the fractionation and determination of natural corticosteroids either from extracts of organs or from biological liquids.

Our interest and previous experience in the analysis of extracts of the adrenal cortex[13-14] have encouraged us to apply this chromatographic method to the qualitative and quantitative analysis of the corticosteroids present in both cortical and human urinary extracts.

The results of our experiment which we report here show that thin-layer chromatography gives excellent separations of corticosteroids and has also some advantages over paper chromatography and column chromatography. These advantages are rapidity of analysis and immediate control of the separation, smaller amounts required, the possibility of using many different reagents for the detection and excellent quantitative results on a microscale.

2. EXPERIMENTAL

Preparation of plates

The Desaga apparatus is used with silica gel G (Merck) as adsorbent. The standard thickness of the layer is 0.250 mm. Plates are either 20 × 20 cm, 15 × 37 cm or 20 × 37 cm. The plates are air dried and heated in an oven at 110°C for 30′, then kept in a desiccator over silica gel.

Development

Glass chambers 18 × 24.5 × 21.5 cm are required for 20 × 20 plates, and a 25 × 40 × 29.5 cm chamber for either 20 × 37 or 15 × 37 plates. With a movable glass support it is possible to change the inclination of plates. With 20 × 37 cm or 15 × 37 cm plates the inclination is adjusted to 20° or to 45°. For 20 × 20 plates the inclination is adjusted to 45°. The solvent is placed in small glass tanks having a capacity of 150 ml, which are situated at the bottom of the development chamber. 100–150 ml of solvent are used in order to submerge an area of 1–1.5 cm of the lower edge of the plate. The two parallel sides of the chamber are lined with two strips of filter paper held in place by metal clips and dipping into the solvent; "complete saturation" is thus attained.

Preparation of samples and application

The pure steroids are dissolved in methanol–chloroform, 1 : 1, at a concentration of 0.5 to 2 mg/ml. The solutions of cortical extracts are prepared by extracting the aqueous solution of the cortical extract with chloroform, as described in detail in one of our preceding papers[13]. The purified chloroform extracts are then dried and the residue taken up in pure ethanol for the quantitative analysis, and in chloroform–methanol, 1 : 1, for both paper and thin-layer chromatography. The tests on thin layers are performed with various quantities, from 5 to 20 µg for the pure steroids and from 100 to 250 µg for the cortical extracts, applying in the latter case up to 100 µl.

Solvent systems

The following solvent systems are used: chloroform–absolute ethanol, (90 : 10), chloroform–90 % methanol, (90 : 10), and cyclohexane–ethyl acetate, (50 : 50).

All solvents used were analytically pure reagents, and if necessary, redistilled.

Detection

Steroids which absorb in the U.V. are rendered visible by using plates incorporating fluorescein (a 0.004 % solution of fluorescein being used for the suspension of silica gel G). This treatment does not hinder the subsequent spraying with tetrazolium blue reagent prepared from 4 ml of BT (0.5% alcoholic solution) and 6 ml of 2 N NaOH. The plates are heated for some minutes at 80°, yielding a pink-violet colour in the presence of ketolic-steroids. Sensitivity 2–3 μg.

Quantitative determination

This is performed by measuring the U.V. absorption of the sample at the wavelength corresponding to its maximum absorption in pure ethanol (suitable for spectrophotometry), measuring if necessary the optical densities from 225 to 260 mμ to identify its spectrum. Alternatively one can use the tetrazolium blue reaction according to NOWACZYNSKY et al.[15], modified as follows: to the sample, containing from 10 to 40 μg of steroids with the —CO—CH$_2$—OH group on the C_{17} atom, in 2 ml of anhydrous and aldehyde-free ethanol, is added 0.2 ml of tetramethylammonium hydroxide (1% in ethanol) and 0.2 ml of tetrazolium blue (0.5% in anhydrous ethanol). After standing 60′ in the dark at 25°C, 1 ml of acetic acid is added and the optical density then measured at the maximum of absorption of the formazane obtained.

It is necessary to prepare a calibration curve, *e.g.* with hydrocortisone, using the above mentioned amounts. For plates treated with fluorescein the absorption in the U.V. cannot be used for quantitative determination, while the tetrazolium blue colorimetric method can be used both with or without fluorescein in the plates. To check the recovery, solutions containing 1 to 2.5 mg/ml in methanol–chloroform 1 : 1 were applied with calibrated pipettes (volumes of 5–10–20 μl and beyond, up to 100 μl). The slides treated with fluorescein permit the direct detection of the $\varDelta^{4-3}$ ketosteroids, while those not treated require parallel chromatography which must be developed using the tetrazolium blue reagent. The spots corresponding to the steroids to be analyzed are removed by scraping from the plate with a small spatula, placed in centrifuge tubes with ground-in stoppers, and extracted with 4–8 ml of ethanol. The adsorbent is separated by centrifugation and the steroids determined in the clear supernatant. For this only the marginal lanes can be utilized. It is preferable, however, to use the device described by MATTHEWS et al.[12], which permits the collection of the adsorbent by suction on a filter glass disc G3 and the direct elution of the steroid through the filter by means of 5–8 ml of ethanol. Both for determination in the U.V. and with tetrazolium blue reagent a blank must be prepared using an area of the plate free from sample. If the absorptions of the blanks are too high in the U.V., it is preferable to use silica gel G washed three times with boiling methanol (ratio 3 : 1, volume/weight) filtered on a G3 disc and dried in an oven.

3. RESULTS AND DISCUSSION

Separations

Table 1 gives the results obtained in the separation of the various steroids with the different solvent systems. In this table various movements are indicated as M_R (relative movement), using cortisone as a reference substance. This value has a

TABLE 1

R_F VALUES OF CORTICOSTEROIDS

Steroid	Relative movement (cortisone reference = 1)		
	System No. 1	System No. 2	System No. 3
(1) THF ($3\alpha,11\beta,17\alpha,21$-tetrahydroxy-pregnane-20-one)	0.33	0.22	0.55
(2) THE ($3\alpha,17\alpha,21$-trihydroxypregnane-11,20-dione)	0.53	0.38	0.59
(3) Cortisone	1.00	1.00	1.00
(4) Hydrocortisone	0.66	0.50	0.75
(5) Corticosterone	1.02	1.19	1.00
(6) Compound S ($17\alpha,21$-dihydroxy-$\triangle^4$-pregnene-3,20-dione)	1.16	1.31	2.30
(7) Compound A (21-hydroxy-$\triangle^4$-pregnene-3,11,20-trione)	1.25	1.60	1.00
(8) Cortexone	1.40	2.00	3.00
(9) Aldosterone	0.85	0.85	0.35
(10) THB ($3\beta,11\beta,21$-trihydroxyallopregnane-20-one)	0.73	0.65	1.00
(11) THA ($3\alpha,21$-dihydroxypregnane-11,20-dione)	0.79	0.91	0.40
(12) 17α-hydroxyprogesterone	1.45	1.52	—
(13) DHF ($11\beta,17\alpha,21$-trihydroxypregnane-3,20-dione)	0.70	0.68	1.30
(14) DHE ($17\alpha,21$-dihydroxypregnane-3,11,20-trione)	1.08	0.92	1.40
(15) DHS ($17\alpha,21$-dihydroxypregnane-3,20-dione)	1.15	1.28	3.60
(16) THS ($3\alpha,17\alpha,21$-trihydroxypregnane-20-one)	0.80	0.59	1.70
(17) Prednisone	0.94	0.82	0.80
(18) Prednisolone	0.62	0.45	0.75
(19) Dexamethasone	0.72	0.51	1.40
(20) Triamcinolone	0.00	0.23	0.10
(21) Dehydroepiandrosterone	—	—	5.10
(22) Androsterone	—	—	5.00
(23) Androstan-3,17-dione	—	—	6.30
(24) Etiocholanolone	—	—	4.00
(25) Etiocholane-3α-hydroxy-11,17-dione	—	—	1.90
(26) Etiocholane-$3\alpha,11\beta$-dihydroxy-17-one	—	—	2.50

better reproducibility than the R_F which is influenced by a large number of variables up to 9, according to BRENNER et al.[16]). The solvent systems (1) and (2) (chloroform with 10% of pure ethanol and chloroform with 10% of 90% methanol) permit the separation of virtually all the steroids present in the adrenal cortex and in urinary extracts.

Figs. 1, 2, 3 and 4 show some of these separations.

It is evident that system (1) allows a better separation of the slow-moving components occurring between the starting point and the hydrocortisone, namely tetra-

References p. 193

hydrocortisol (THF), tetrahydrocortisone (THE) and some of the minor components of the cortical extract which move in this area[14]. The system (2) gives, on the other hand, better separation of the components faster than cortisone and permits the separation of cortisone and corticosterone which move with same speed in system (1). With both systems the separation of aldosterone from cortisone and hydrocortisone is obtained, and also from tetrahydrocorticosterone (THB) and tetrahydro-11-dehydrocorticosterone (THA) which are present in cortical and urinary extracts. The sequence of the components is a little different, being aldosterone–THA–THB in system (1) and THA–aldosterone–THB in system (2). These systems among the many proposed in literature and others designed and tried by us, have proven themselves

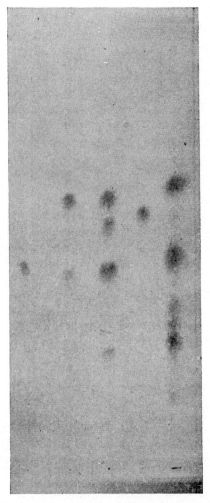

Fig. 1. Chromatogram with solvent 1. Plate 15 × 37 cm. From left to right and from below to above: (1) hydrocortisone — cortisone. (2) corticosterone — compound A. (3) hydrocortisone — cortisone + corticosterone — compound S — compound A. (4) compound S. (5) cortical extract

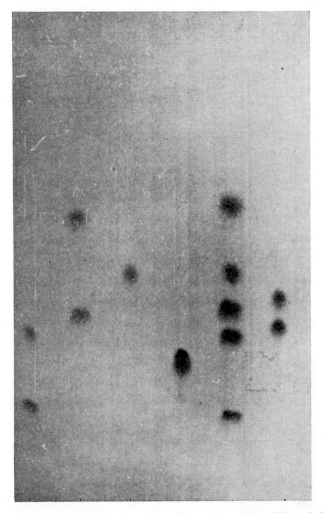

Fig. 2. Chromatogram with solvent 2. Plate 20 × 37 cm. From left and from below: (1) Hydrocortisone — cortisone. (2) corticosterone — compound A. (3) compound S. (4) aldosterone. (5) hydrocortisone — cortisone — corticosterone — compound S — compound A. (6) cortisone — corticosterone.

best suited to the fractionation using only one chromatogram of the whole of the corticosteroids present in a cortical or urinary extract, and permits their subsequent quantitative determination. Longer plates than the standard ones (that is 20 × 37 cm instead of 20 × 20 cm), are more convenient because the improved separation of the components in a longer run allows a precise isolation of the single spots.

The running time of the solvent is about 2 h for this type of plate, with an inclination of 20° and for a length of about 30 cm. It rises to 2 h 45′ with an inclination of 45°.

A steeper inclination, even if small, produces an increase in the R_F values and there-

References p. 193

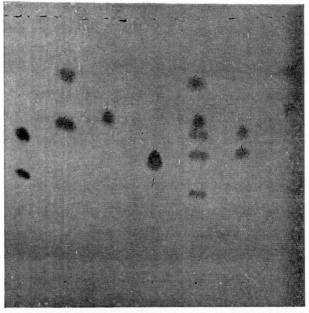

Fig. 3. Chromatogram with solvent 2. Plate 20 × 20 cm. (1)–(6) (from left) as in Fig. 2. (7): compound S — compound A.

Fig. 4. Chromatogram with solvent 2. Plate 5 × 37 cm. From below: THF — THE — hydrocortisone — THB — aldosterone — cortisone — corticosterone — compound S — compound A.

fore an effect similar to that of the polarity increase of the solvent. The values of M_R do not change perceptibly on passing from a 20 × 37 cm plate to one of 20 × 20 cm. The latter allows a more rapid orientating evaluation of a complex mixture or a quantitative determination on a limited number of components.

System (3) is convenient for analyzing corticosteroids of low polarity and in the presence of 17-ketosteroids, which are satisfactorily separated from each other (Fig. 5).

Quantitative determination

Table 2 shows the results obtained in the recovery tests performed with quantities from 12.5 to 100 μg of hydrocortisone. Determinations were performed with the spectrophotometric method in ethanolic solutions at 242 mμ and with the tetrazolium blue colorimetric method on samples submitted to chromatography or simply placed on plates. In every case blanks were performed by extracting silica gel G. Silica gel G stocks in our possession have generally given very low blank-values, even without preliminary washing. The average O.D. value (40 determinations) at 242 mμ is 0.011 for a 2.5 × 2.5 cm area in 8 ml of ethanol. The tetrazolium blue reaction at 495 mμ is of about the same magnitude.

Table 2 also shows that recoveries are not perceptibly different between the developed samples and those not developed. This indicates that during development

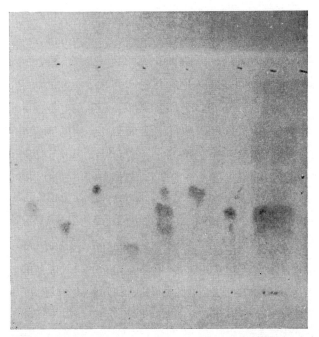

Fig. 5. Chromatogram with solvent 3. Plate 20 × 20 cm. From left and from below: (1) Hydrocortisone + cortisone. (2) corticosterone + compound A. (3) compound S. (4) aldosterone. (5) corticosterone + compound A — hydrocortisone + cortisone — compound S. (6) desoxycorticosterone. (7) corticosterone — cortisone. (8) cortical extract.

References p. 193

TABLE 2
QUANTITATIVE DETERMINATION OF CORTICOSTEROIDS

Quantity of steroid spotted on plate	Quantity recovered (average of 7 to 12 determinations)			
	Without development		After development	
	Colorimetric	Spectrophotometric	Colorimetric	Spectrophotometric
Hydrocortisone 12.5 µg	11.11 µg (88.2%) $\delta = 0.84$	10.65 µg (85.2%) $\delta = 0.49$	10.05 µg (80.5%) $\delta = 0.26$	10.42 µg (83.4%) $\delta = 0.35$
Hydrocortisone 25 µg	22.27 µg (89.1%) $\delta = 0.29$	23.65 µg (94.6%) $\delta = 0.11$	23.26 µg (93.0%) $\delta = 0.70$	23.98 µg (95.92%) $\delta = 0.64$
Hydrocortisone 50 µg	43.78 µg (87.56%) $\delta = 0.94$	46.98 µg (93.96%) $\delta = 0.16$	44.87 µg (89.74%) $\delta = 0.94$	45.55 µg (91.10%) $\delta = 0.55$
Hydrocortisone 100 µg	89.72 µg (89.72%) $\delta = 1.25$	92.08 µg (92.08%) $\delta = 3.01$	87.53 µg (87.53%) $\delta = 1.92$	91.50 µg (91.50%) $\delta = 1.98$

$\delta = \sqrt{\dfrac{\Sigma\,(s)^2}{n(n-1)}}$; s = deviation of individual values from the mean, n = number of determinations.

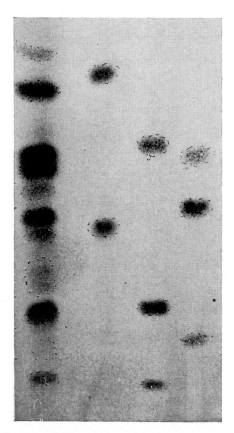

Fig. 6. Chromatogram with solvent 2. From right to left and from below: (1) THE — cortisone — corticosterone. (2) THF — hydrocortisone — compound S. (3) aldosterone — compound A. (4 cortical extract (a pharmaceutical preparation). The fastest spot is a trace of desoxycorticosterone

the band migrates in an almost compact manner. No appreciable differences were observed whether the areas were removed by scraping or by aspiration.

Table 3 shows the results of recovery tests performed on samples of cortical extract and pure steroid mixtures. Previously, cortical extracts were analyzed by paper chromatography, using the technique described by us[13] and their composition is shown on the same table. Results agree, especially for the major components. One can observe how even minor components can be separated and determined by means of only one chromatogram. Compound S moves in paper chromatography together with corticosterone and requires further chromatography with the E_2B system according to EBERLEIN AND BONGIOVANNI[17] on the eluted corticosterone spot, while the isolation of it as a single component can be performed by the thin-layer method (Figs. 2, 6). As described in a preceding note[13] two paper chromatograms (Bush C and Bush B1) are required to get a convenient separation of principal components of the cortical extract and a succession of two developments is required in order to measure aldosterone in

References p. 193

Sample	System	Quantity of sample spotted, calculated as hydrocortisone (BT evaluation)	% of single fraction re			
			THF fraction	THE fraction	Hydro-cortisone fraction	fr
(1) Synthetic mixture of hydrocortisone, cortisone, corticosterone, compound A	2	4 × 25 µg	—	—	20.5	
(2) Synthetic mixture of hydrocortisone, cortisone, corticosterone, compound A	1	4 × 25 µg	—	—	21.9	
(3) Synthetic mixture of hydrocortisone, cortisone, compound A, compound S	2	4 × 25 µg	—	—	21.6	
(4) Bovine extract	2	219 µg	3.58	11.70	13.20	
(5) Idem	P	190 µg	7.78	15.90	18.0	
(6) Cortical extract for pharmaceutical use	2	229 µg	4.12	8.92	28.95	
(7) Idem	P	96.5 µg	5.23	9.62	24.05	
(8) Urinary extract	2	150 µg	7.03	17.75	5.82	

Systems: 1. chloroform : ethanol = 90 : 10; 2. chloroform : methanol : water = 90 : 9 : 1;
3. P : paper chromatography; (°) in paper chromatography compound S and corticosterone

a semiquantitative manner. With one thin-layer chromatographic run using 200–150 µg of a cortical extract one can directly determine aldosterone on the respective spot together with the other components of the extract (Fig. 7), finally evaluating its purity by means of paper chromatography, according to NEHER et al.[18]. Aldosterone isolated from the corresponding spot behaves as a single component, with characteristics of movement similar to those of a pure aldosterone sample. Also shown in Table 3 is an example of human urine extract analysis, obtained with chloroform after β-glucuronidase hydrolysis according to GLENN AND NELSON[19], without chro-

...-ne ...on	Cortisone fraction	Cortico-sterone fraction	Compound S fraction	Compound A fraction	Other fraction	Other fraction	Total recovery μg	% recovery
	20.9	21.1	—	23.9	—	—	86.4	86.4
	44.8	(°°)	—	23.9	—	—	90.6	90.6
	20.2	—	22.3	21.7	—	—	85.8	85.8
	13.60	7.8	23.2	19.0	—	—	210.1	96.0
	15.40	24.6	(°)	16.0	—	—	173.0	91.0
	15.50	15.85	3.59	12.26	3.10(.)	—	225.4	98.5
	15.65	11.65	(°)	10.35	—	3.52(:)	83.1	86.2
	19.05	7.0	11.4	6.47	3.45(.)	—	131.5	88.0

move together;   (:) fraction less moving than THF;   (.) fraction between compound S and compound A;   (°°) in system 1 cortisone and corticosterone move together.

matographic fractionation on Florisil. The rough extract was purified from 17-ketosteroids and non-specific reducing agents by previous thin-layer chromatography in chloroform with 1% pure ethanol.

The spot corresponding to corticosteroids (0–2.6 cm) is eluted, concentrated, and submitted to chromatography[2]. A regular separation of the various components is obtained with well isolated bands which can be eluted and analyzed by the tetrazolium blue reaction (Fig. 8). The results of a more detailed application of this technique to the analysis of cortical and urinary extracts will be published separately.

References p. 193

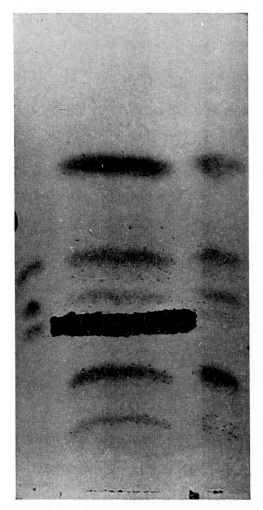

Fig. 7. Chromatogram with solvent 2. From left and from below: (1) aldosterone — cortisone — corticosterone. (2) cortical extract (600 μg were deposited as a line — the aldosterone was removed and the remaining zones detected with BT). (3) cortical extract (all zones detected with BT).

4. SUMMARY

Some solvent systems and experimental conditions are described for the separation and determination of the principal corticosteroids present in cortical and urinary extracts with a single thin-layer chromatogram. After the separation, the steroids are quantitatively extracted from the respective areas and determined by colorimetry with tetrazolium blue or by the measurement of U.V. absorption.

The advantages of the method described, are:

(1) Rapidity of analysis: only a 2 h 30′ development is needed to fractionate a cortical extract into its components.

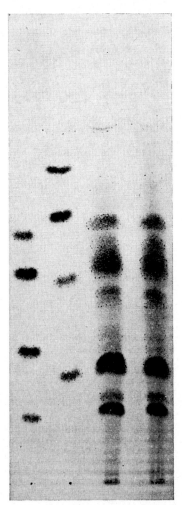

Fig. 8. Chromatogram with solvent 2. From left and from below: (1) THF — hydrocortisone — cortisone — corticosterone. (2) THE — aldosterone — compound S — compound A. (3) urine extract. (4) urine extract:

(2) Less substance can be used in an analysis; 100 to 250 µg of a corticoid mixture is sufficient to perform the fractionation and determination of 8–9 components with satisfactory recovery.

(3) Immediate control of the separation and possibility of using more than one reagent for the detection of the various components.

REFERENCES

1. E. STAHL, *Dünnschicht-Chromatographie*, Springer, Berlin, 1962.
2. M. BARBIER, H. JÄGER, H. TOBIAS AND E. WYSS, *Helv. Chim. Acta*, 41 (1959) 2440.
3. R. NEHER AND A. WETTSTEIN, *Helv. Chim. Acta*, 43 (1960) 1628.
4. L. STARKA AND J. MALIKOVA, *J. Endocrinol.*, 22 (1961) 215.

5. B. P. LISBOA AND E. DICZFALUSY, *Acta Endocrinol.*, 40 (1962) 60.
6. H. STRUCK, *Mikrochim. Acta*, (1961) 634.
7. V. CÉRNY, J. JOSKA AND L. LABLER, *Collection Czech. Chem. Commun.*, 26 (1961) 1658.
8. S. HERMANEK, W. SCHWARZ AND Z. CÉKAN, *Collection Czech. Chem. Commun.*, 26 (1961) 1669.
9. O. ADAMEC, J. MATIS AND M. GALVANEK, *Lancet*, (1962-I) 81.
10. B. P. LISBOA, *Congr. Intern. Steroidi Ormonali*, Milano, May 1962.
11. R. D. BENNET AND E. HEFTMANN, *J. Chromatog.*, 9 (1962) 348.
12. J. S. MATTHEWS, A. L. PEREDA AND A. P. AGUILERA, *J. Chromatog.*, 9 (1962) 331.
13. G. CAVINA, E. CINGOLANI AND L. TENTORI, *Farmaco (Pavia), Ed. Prat.*, 16 (1961) 3.
14. G. CAVINA, *Boll. SIBS*, 37 (1961) 1818.
15. W. NOWACZYNSKI, M. GOLDNER AND J. GENEST, *J. Lab. Clin. Med.*, 45 (1955) 818.
16. M. BRENNER, A. NIEDERWIESER, G. PATAKI AND A. R. FAHMY, *Experientia*, 18 (1962) 101.
17. W. R. EBERLEIN AND A. M. BONGIOVANNI, *Arch. Biochem. Biophys.*, 59 (1955) 90.
18. R. NEHER AND A. WETTSTEIN, *J. Clin. Invest.*, 35 (1955) 800.
19. E. M. GLENN AND D. H. NELSON, *J. Clin. Endocrinol. Metab.*, 13 (1953) 911.

THIN-LAYER CHROMATOGRAPHY AND THE DETECTION OF STILBOESTROL

CESARE BONINO

Pharmaceutical Chemistry Institute, University of Bologna (Italy)

SUMMARY

Since the addition of stilboestrol (= diethylstilbestrol or a,a'-diethyl-4,4'-stilbenediol) to feed was forbidden, owing to its cancerogenic properties, its detection is of special interest.

The official method of analysis of stilboestrol[1] is based on the U.V.-irradiation of an extract of the sample followed by spectrophotometric measurement of the absorbance of the yellow colour developed.

Fig. 1. Spot of stilboestrol after U.V.-irradiation of the layer for 2 min.

A simple, rapid method would be useful for the detection of stilboestrol. For the paper chromatography of stilboestrol, reversed-phase chromatography was employed, using propylene glycol/toluene as solvent system and the ferric chloride–ferricyanide reagent for detection[2]. This method is slow and non-specific.

We have achieved good results with thin-layer chromatography on silica gel, using a chloroform–ethanol mixture as solvent system and U.V.-irradiation of the layers for the detection of stilboestrol.

After a microevaporation of the chloroform extracts of the acidified samples, with pure nitrogen at 35° C in small centrifuge tubes, the chromatography of the residues was carried out on silica-gel thin layers with chloroform–ethanol 8 : 2 (v/v). The layers were then irradiated for 2 min with U.V.-light from a 0.5-m distant 500-W lamp. Stilboestrol gave faint yellow very fluorescent spots (Fig. 1). The layers were then examined in U.V. Wood light: 0.1 γ of stilboestrol could still be detected. A rough determination was also possible — the method is rapid and quite specific.

Thin-layer chromatography can replace reversed-phase paper chromatography with great advantage in this case.

REFERENCES

1. *Official Methods of Analysis of the Association of Official Agricultural Chemists*, IX ed., 1960.
2. P. H. JELLINEK, Paper chromatography of estrogens, *Nature*, 171 (1953) 750.

THE ANALYSIS OF MIXTURES OF ANIMAL AND VEGETABLE FATS

IV. SEPARATION OF STEROL ACETATES BY REVERSED-PHASE THIN-LAYER CHROMATOGRAPHY*

J. W. COPIUS PEEREBOOM

Government Dairy Station, Leiden (The Netherlands)

1. INTRODUCTION

While studying the separation of sterols and sterol acetates by reversed-phase thin-layer chromatography (rev-phase T.L.C.), we found a correct choice of the stationary phase to be of prime importance. When testing stationary phases such as paraffin and silicone oils, difficulties were encountered in the detection of minor components in certain sterol mixtures (down to 1%). The volatile stationary phase undecane** (b.p. 190–220° C), is therefore preferred, since it can be removed almost completely from the chromatoplate after the development procedure.

Mixtures of acetic acid and water were first used as mobile phases[1] but the long time of development (*ca.* 5 h) using these mixtures was somewhat tiresome. Shorter times (down to 1.5 h) can be achieved using acetic acid–acetonitrile mixtures: this was successfully employed in the separation of triglycerides by KAUFMANN *et al.* (see *e.g.* Ref. 2).

Up to the present, the clearest separations in the group cholesterol, β-sitosterol and related sterols, have been achieved using the system undecane/acetic acid–acetonitrile (1 : 3)[7].

Briefly, the rev-phase T.L.C. of sterol acetates was carried out as follows. A thin layer of kieselgur G (Merck) is spread upon a glass plate according to STAHL. After activating, the chromatoplate is dipped into a 10% solution of undecane in light petroleum. The solvent, and also a certain portion of the undecane, is evaporated over a standard period (usually 80 min). The plate is modelled with a templet according to the principle of MATTHIAS. This so-called wedged-tip technique enables separations to be made which would be impracticable when using the normal one-dimensional techniques. 20–40 μg of sterol acetates is generally spotted. The plate is developed in a completely saturated small chromatographic vessel in 1.5–2 h at a temperature of about 23° C. The chromatoplate is dried, then sprayed with a phosphomolybdic acid solution and heated.

---

* For Parts I, II and III of this series, see Refs. 3, 4 and 1.
** Available from J. Haltermann, Hamburg.

2. DISCUSSION

General system

In this procedure many of the experimental conditions appear to be rather critical. The detailed procedure given in the experimental part should therefore be strictly followed. With this procedure cholesterol and the phytosterols from soybean oil are separated into three bands *viz.* band 1 containing β-sitosterol, band 2: stigmasterol and C28-phytosterols, and band 3: cholesterol (Fig. 1, spot 3).

The R_S (S = cholesterol) and R_M values of some sterol acetates in this system are given in Table 1.

TABLE 1

R_S* AND R_M VALUES OF SOME STEROL ACETATES IN THE SYSTEM
UNDECANE/ACETIC ACID–ACETONITRILE (1 : 3)

Compound	Shorthand designation	R_S value	R_M value
Cholesterol	FC 27	$\equiv 1.0$	0.410
Stigmasterol	FC 29 F**	0.91	0.465
β-Sitosterol	FC 29	0.83	0.520
Brassicasterol	FC 28 F	1.07	0.368
Ergosterol	2FC 28 F	1.19	0.302
7-Dehydrocholesterol	2FC 27	1.16	0.317
Dihydrocholesterol	C 27	0.89	0.479
Dihydro-β-sitosterol	C 29	0.73	0.590

* S = cholesterol.
** This formula means a C 29 sterol skeleton with one double bond in the nucleus and one in the side chain of the molecule respectively.

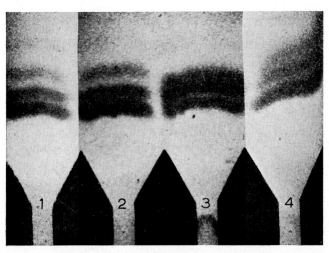

Fig. 1. Separation of some phytosterol acetate mixtures by rev-phase T.L.C. in the system undecane/acetic acid–acetonitrile (1 : 3). Time of run: 1.5 h. Length of run: 19 cm. Detection: phosphomolybdic acid. Spotted amount: 30 μg of the phytosterol acetate mixtures from: Spot 1 = sesame oil. Spot 2 = crude peanut oil. Spot 3 = a cholesterol–soybean oil phytosterol (2 : 8) reference mixture. Spot 4 = refined palm kernel oil.

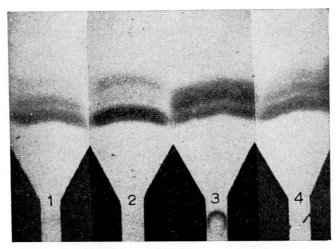

Fig. 2. Separation of some phytosterol acetate mixtures by rev-phase T.L.C. in the system undecane/acetic acid–acetonitrile (1 : 3). Time of run 1.5 h. Length of run: 19 cm. Detection: phosphomolybdic acid. Spotted amount: 30 μg of the phytosterol acetate mixtures from: Spot 1 = sunflower oil. Spot 2 = olive oil. Spot 3 = a cholesterol–soybean oil phytosterol (2 : 8) reference mixture. Spot 4 = refined palm kernel oil.

On the whole the separations obtained in this system of rev-phase T.L.C. are analogous to those of reversed-phase paper chromatography. In the paper chromatographic system paraffin/acetic acid–water (84 : 16)[3] we have previously found a relationship between the R_M value and the difference between the number of carbon atoms and double bonds of the sterol molecule[4]. On account of this relationship the sterols were classified into several so-called critical pairs. The various bands are constituted by the following sterol types:

band 4:   2FC 28 F; 2FC 27; FC 27 F
band 3:   FC 28 F; FC 27
band 2: FC 29 F; FC 28 ; C 27
band 1: FC 29
band 0: C 29

In the rev-phase T.L.C. of sterol acetates we have found nearly the same relationship. However, in contrast to the situation in reversed-phase paper chromatography, in rev-phase T.L.C. there exists a small but marked difference in migration rate between cholesterol (FC 27) and its critical-pair partner FC 28 F (the so-called FC 28 F sterols), as is shown in Figs. 1 and 2.

A similar difference in the migration rates of the critical-pair triglycerides PPP, PPO, POO and OOO* after multiple development in the system paraffin/acetone–acetonitrile (8 : 2) has been reported by KAUFMANN et al.[2].

In addition to this difference in R_S value, cholesterol and the "FC 28 F sterols" may also show sufficiently differing colours upon spraying with certain specific colour reagents e.g. bismuth(III) chloride (33% in 96% ethanol) and antimony(III) chloride.

References p. 204

Figs. 1 and 2 present the difference in R_S value of cholesterol (Fig. 1, spot 3) and the "FC 28 F phytosterols" from peanut oil (Fig. 1, spot 2), sunflower oil (Fig. 2, spot 1), and sesame oil (Fig. 1, spot 1). Many other types of edible oils and fats present a sterol pattern similar to that of e.g. sunflower oil, with certain deviations in the quantitative composition.

However, the phytosterol composition of some edible, vegetable oils shows peculiar details. Thus, in the chromatogram of the sterol acetates from sesame oil a very faint fourth band is revealed, possibly due to a sterol of the ergosterol–zymosterol critical pair. The phytosterols from olive oil show a band with nearly the same migration rate as cholesterol. This band might possibly be caused by a sterol of the provitamin D type, e.g. 3 FC 28 F or 3 FC 27 (vide Fig. 2, spot 2).

Furthermore, the chromatograms of the sterol mixtures from crude palm oil, refined palm oil, and especially those from refined palm kernel oil, reveal a small band due to an unidentified sterol in the cholesterol position (Fig. 1, spot 4 and Fig. 2, spot 4). As we have previously reported[3], in reversed-phase paper chromatography these sterol mixtures reveal quite a "normal" third band at the position of the cholesterol–brassicasterol critical pair.

However, the results of the separations in rev-phase. T.L.C. indicate that this peculiar sterol is *not* identical with a FC 28 F sterol or brassicasterol. Upon spraying with specific colour reagents e.g. bismuth(III)chloride and antimony(III)chloride this palm kernel sterol shows a colour nearly the same as that of cholesterol, but quite different from that given by the FC 28 F "brassicasterol-like" phytosterols.

These results are in agreement with the gas chromatographic analysis of palm oil phytosterols, carried out by RECOURT[5]. This author has reported the occurrence of a special sterol in the palm oil sterol mixture revealing a retention time quite similar to that of cholesterol.

Analysing some samples of crude and refined palm kernel oil, we found the percentage of this special sterol in the total sterol mixture to be increased by the refining process. We therefore may suggest that most probably this third band of palm kernel oil sterol acetates does not consist of any cholesterol from natural origin but contains a special hitherto unidentified phytosterol type.

Although the composition of all phytosterol mixtures is subject to a natural variability, the above-mentioned T.L.C. sterol patterns may represent an average sterol composition of the respective oils and fats.

In crude tall oil and also in several hardened vegetable oils a so-called "zero band", due to the presence of hydrogenated β- and γ-sitosterols (C 29 sterols) is shown on the chromatoplate (Fig. 3, spot 4).

"Bromine system"

In general, critical pairs of fatty acids and triglycerides are resolved by additional *chemical* procedures e.g. by hydrogenation, bromination or oxidation of the un-

---

* PPP: tripalmitin, OOO: triolein etc.

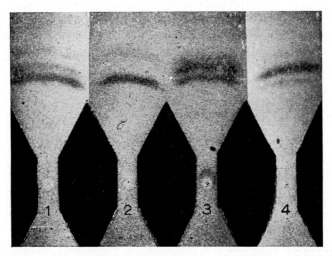

Fig. 3. Separation of some phytosterol acetate mixtures by rev-phase T.L.C. in the system undecane/acetic acid–acetonitrile (1 : 3). Time of run 1.5 h. Length of run: 20 cm. Detection: phosphomolybdic acid. Spotted amount 20 µg of the phytosterol acetates from: Spot 1 = refined rape seed oil. Spot 2 = refined almond oil. Spot 3 = a cholesterol–soybean oil phytosterol (2 : 8) reference mixture. Spot 4 = crude tall oil.

saturated members of the series. In our study[7] we have applied the bromination procedure elaborated by KAUFMANN AND MAKUS[6]. In this simple and elegant procedure we can use the same solvent system as that used for separating the sterol acetates. A quantity of 0.5% of bromine is added to the mobile phase immediately prior to starting the development procedure.

In this bromine containing mobile phase the sterol acetates are brominated quanti-

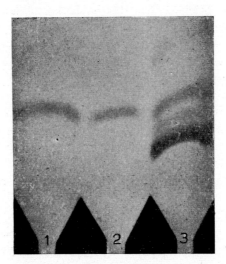

Fig. 4. Separation of sterol acetates in the "bromine system" viz. undecane/acetic acid-acetonitrile (1 : 3) + 0.5% of bromine[7]. Time of run: 2 h. Length of run: 18 cm. Detection: heating and spraying with antimony(III) chloride (50% in acetic acid). Spot 1 = stigmasterol acetate. Spot 2 = cholesterol acetate. Spot 3 = 30 µg of coconut phytosterol acetates.

References p. 204

tatively during the development procedure. After development, the chromatoplate is dried, heated and afterwards sprayed with a 50% solution of antimony(III)chloride in acetic acid. The sterol acetate bromides are revealed as bright blue bands.

The phytosterol acetates from coconut fat are by this means separated into four bands; band 1 containing β-sitosterol, band 2: C 28-phytosterols such as campesterol, band 3: stigmasterol, and band 4: the above-mentioned FC 28 F phytosterols (Fig. 4, spot 3). As compared with the "normal system" the pattern of sterol bands is thus considerably changed. The R_S and R_M values of some sterols are given in Table 2.

TABLE 2

R_S AND R_M VALUES OF SOME STEROL ACETATES IN THE "BROMINE SYSTEM" *viz.* UNDECANE/ACETIC ACID–ACETONITRILE (1 : 3) + 0.5 % OF BROMINE

Compound	Shorthand designation	R_S value	R_M value
Cholesterol	FC 27	≡ 1.0	0.368
Stigmasterol	FC 29 F	1.06	0.331
β-Sitosterol	FC 29	0.84	0.472
C 28-phytosterols	FC 28	0.95	0.399
"FC 28 F phytosterols"	FC 28 F	1.3–1.4	0.19–0.17
Dihydrocholesterol	C 27	0.85	0.465
Ergosterol	2FC 28 F	front	—
7-Dehydrocholesterol	2FC 27	front	—
Lanosterol	FC 30 F	front	—

In this bromine system stigmasterol now has about the same migration rate as cholesterol (Fig. 4, spots 1 and 2). The migration rate of brominated "FC 28 F sterol acetate" likewise is much higher than that of cholesterol acetate, thus enabling an unambiguous differentiation between both "critical-pair" sterols. Dihydrocholesterol acetate contains no double bond and thus cannot be brominated. It is therefore not revealed by a bright blue band, but only shows a faint yellow band at the same position as in the "normal system".

The most common naturally occurring type of sterol, *viz.*, that with a $\triangle^{5(6)}$ double bond in the nucleus, is quite stable under the conditions of the bromine system. However, sterols with conjugated double bonds in the nucleus such as ergosterol and 7-dehydrocholesterol are completely decomposed by the bromine and only show a large blue spot near the solvent front. Remarkably, sterols having non-conjugated nuclear double bonds in *other positions* than $\triangle^{5(6)}$ are similarly decomposed. Thus all $\triangle^7$-sterols such as *e.g.* $\triangle^7$-stigmastenol (F7 C29), α-spinasterol (F7 C29 F), $\triangle^7$-cholestenol (or lathosterol, F7 C27), and also sterols like zymosterol [$\triangle^{8(9), 24(25)}$-cholestadiene-3β-ol] do not show any band on the chromatoplate exept the spot near the solvent front. The tetracyclic triterpenoid alcohols lanosterol, agnosterol, dihydroagnosterol etc. also behave similarly.

This difference in behaviour of $\triangle^{5(6)}$-sterols and of other sterol types may be of use in the structural analysis of special naturally occurring sterol mixtures. Rev-phase

T.L.C. is also very suitable for the analysis of mixtures of vegetable and animal fats. In this way the purity of fats of animal origin, *e.g.* butter fat, can be investigated, while on the other hand admixtures of cheap animal fats to pure vegetable fats, *e.g.* margarines labeled as "pure vegetable", can be detected.

3. EXPERIMENTAL PART[7]

Separation of sterol acetates in the system undecane/acetic acid–acetonitrile (1 : 3)

Glass plates of 14 × 24 cm are coated with a layer of kieselgur G, Merck, according to the procedure of STAHL. After activation for 15 min at 100° C, the resulting layer is 0.24–0.28 mm thick. After cooling to room temperature, the plate is taken between thumb and forefinger of both hands (wearing rubber gloves) and is dipped carefully into a shallow rectangular vessel containing a 10% solution of undecane (previously saturated with the mobile phase) in petroleum ether (b.p. 40–60°). After some seconds the plate is removed from the solution and is held upside down for some seconds. The plate is then allowed to rest in a horizontal position for a standard time — usually 80 min — at room temperature. During this period hexagonal pieces (23 × 20 mm) are removed from the layer with a brush by using an appropiate templet. In this way four wedged-tip chromatograms are modelled on the chromatoplate. The distance from the centre A of the 8 mm wide "bridges" to the bottom of the plate is 40 mm. An amount of 0.02–0.03 ml of a 0.1% ethereal solution of the sterol acetates is spotted on A by means of a micropipette.

After the evaporation period of 80 min, the plate is developed with the mobile phase *viz.* acetic acid–acetonitrile (1 : 3), in a completely saturated chromatographic jar. Two days previously, this solvent mixture had been shaken vigorously in a separatory funnel with a sufficient amount of undecane. This mixture is kept during 16 h at a constant temperature of 22–23° C. The two layers are then separated. The undecane layer is dissolved in petroleum ether (10%); this solution is used for the impregnation procedure. The chromatographic vessel is supplied with filter paper at the walls. After introducing a sufficient quantity of the undecane-saturated acetic acid–acetonitrile layer, the vessel is equilibrated for 24 h at a temperature of 22–23° C.

During the development procedure the mobile phase will ascend to a height of some 20 cm in only 1.5–2 h. The plate is then dried in air for 2–4 h and afterwards for 45 min at 100° C. The plate is sprayed with a 20% ethanolic solution of phosphomolybdic acid, Merck (the reagent should not come into contact with a metal object such as a spatula). After spraying, the plate is heated for 5–10 min at 100° C till the bands are coloured to the maximum intensity, but avoiding excessive darkening of the background. After being used once, the solvent mixture should be discarded and the vessel should be cleaned thoroughly.

Bromine system[7]

The procedure described above for preparing the plates is used throughout. The temperature for the development procedure should preferably be 18–20° C. 0.5% of

References p. 204

bromine is introduced into the vessel just before starting the development procedure. After 1.5–2 h the development process is discontinued and the plate is dried for 2 h in air and then for about 10 min at 100° C. At this phase of the detection process the sterols are generally already coloured and form blue bands. To increase the colour intensity, the warm plate is afterwards sprayed with a 50 % solution of antimony(III) chloride in acetic acid or with an acetic anhydride–sulphuric acid 50 % (1 : 2) mixture. The sterol bands then become bright blue.

REFERENCES

1. J. W. COPIUS PEEREBOOM AND H. W. BEEKES, *J. Chromatog.*, 9 (1962) 316.
2. H. P. KAUFMANN AND B. DAS, *Fette, Seifen, Anstrichmittel*, 64 (1962) 214.
3. J. W. COPIUS PEEREBOOM AND J. B. ROOS, *Fette, Seifen, Anstrichmittel*, 62 (1960) 91.
4. J. W. COPIUS PEEREBOOM, J. B. ROOS AND H. W. BEEKES, *J. Chromatog.*, 5 (1961) 500.
5. J. H. RECOURT AND R. K. BEERTHUIS, *Fette, Seifen, Anstrichmittel*, 65 (1963) 619.
6. H. P. KAUFMANN, Z. MAKUS AND T. H. KHOE, *Fette, Seifen, Anstrichmittel*, 64 (1962) 1.
7. J. W. COPIUS PEEREBOOM, *Chromatographic Sterol Analysis*, Pudoc, Wageningen, 1963.

APPLICATIONS OF THIN-LAYER CHROMATOGRAPHY ON SEPHADEX TO THE STUDY OF PROTEINS

P. FASELLA, A. GIARTOSIO AND C. TURANO

*Istituto di Chimica Biologica, Università di Roma, e
Centro di Enzimologia del Consiglio Nazionale delle Ricerche (Italy)*

1. INTRODUCTION

Chromatography on Sephadex columns is currently employed for the separation of substances of different molecular size[1]. Chromatography on thin layers of Sephadex G 25 and G 75 has been employed by DETERMANN[2] and by JOHANSSON AND RYMO[3].

In the present work a technique is described for the separation of proteins and peptides of molecular weight ranging from 200 to 2 millions on thin layers of Sephadex. Compared with column chromatography, the present technique offers considerable advantages in rapidity, sensitivity and cost. In comparison with the previously described methods of thin-layer chromatography on Sephadex, the present technique considerably simplifies the preparation of the layers and avoids the use of costly special equipment. The present investigation, moreover, has been extended to the use of low cross-linkage products (Sephadex G 100 and G 200), which give the best separations with proteins. By devising a new procedure for the detection of the separated substances, and by increasing the thickness of the layer, chromatography on Sephadex has also been adapted to micro-preparative purposes. Among analytical applications, the quick determination of the rough molecular weight of proteins deserves special mention. This problem has been systematically investigated in the present work using Sephadex G 100 or G 200. Thin layers of Sephadex have also been employed here for bi-dimensional separations, using chromatography by gel filtration in one direction and electrophoresis in the other. The latter procedure proved particularly helpful in the characterization of the peptides obtained by partial digestion of proteins.

2. EXPERIMENTAL

Materials

Sephadex G 25 (fine), G 100 and G 200 were commercial samples from Pharmacia AB. Solutions of peptides from adult human hemoglobin were obtained as described by INGRAM[4]. Glutamic-aspartic aminotransferase was prepared according to LIS[5]. Hemocyanin was kindly given by Dr. NARDI, of the Stazione Zoologica, Naples.

Bovine pancreas chymotrypsin was a commercial sample of Fluka AG. Bovine serum albumin was a commercial sample from Sigma, Inc. Adult human hemoglobin was prepared according to Rossi Fanelli et al.[6]. All other reagents were commercial samples "pro analysi" from Merck AG.

Methods

Sephadex was suspended in the desired buffer or solvent (50–100 ml of buffer per g of dry Sephadex). After stirring for 30 min the gel was allowed to sediment and the supernatant was decanted. The procedure was repeated 5–6 times at intervals of time long enough to leave the Sephadex in contact with the buffer for at least 48 h in the case of Sephadex G 25 and at least 72 h in the case of G 100 and G 200. For the last sedimentation a graduated cylinder was used so that it was possible to determine the volume occupied by the sedimented Sephadex. Sephadex was then homogeneously suspended in 2 volumes of buffer and the suspension was poured into a flat, square container (made of plastic or glass) with vertical walls and smooth bottom. Two types of container were employed, the sizes of which were respectively 30 cm × 20 cm × 3 cm and 40 cm × 30 cm × 3 cm. The volume of Sephadex suspension poured into each container was calculated to give a Sephadex layer about 1 mm thick. The container was placed in a horizontal position (checked with a spirit level) and the gel was allowed to sediment homogeneously. After the gel had sedimented, the supernatant was removed by absorption on sheets of filter paper as indicated in Fig. 1. The filter paper sheets were then removed and the Sephadex appeared as a homogeneous, humid, relatively stable film on the bottom of the container. The Sephadex film can be stored for several days before use, provided that the container is covered with a glass or plastic plate to prevent drying by evaporation.

Chromatography by molecular filtration

The container with the Sephadex film was arranged as shown in Fig. 2. In the system illustrated in this figure, the fluid flows regularly through the Sephadex layer from the upper to the lower end. The system is then ready to receive the material to be analyzed.

0.01–0.05 ml of a 0.5–5% protein or peptide solution were spotted on the gel at the upper end of the layer. In uni-dimensional separations it was possible to analyze several samples simultaneously. The various samples were spotted at intervals of at least 3 cm one from the other along an imaginary line, normal to the flow of the fluid and passing about 1 cm below the upper end of the layer. In the case of bi-

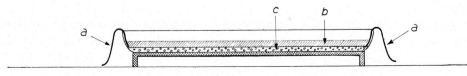

Fig. 1. (a) Filter paper; (b) supernatant fluid; (c) sedimented Sephadex.

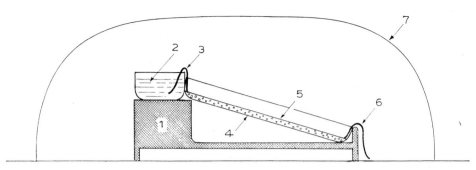

Fig. 2. (1) Supporting system; (2) Trough containing the buffer or solvent used for the development of the chromatogram: the level of the fluid in the trough must be about 1 cm above the upper end of the Sephadex layer; (3) Sheet of Whatman No. 1 paper for the transport of fluid from the trough to the Sephadex layer. One end of the sheet is immersed (2–3 cm) in the fluid contained in the trough, the opposite end lies on the upper end of the Sephadex layer; (4) Container with Sephadex layer; the container should be slanted so as to form a 15–20° angle with the horizontal; (5) Sephadex layer; (6) 3 Sheets of Whatman No. 1 filter paper (for the removal of the fluid at the lower end of the Sephadex layer); (7) Glass or plastic bell for protecting the system and for preventing the excessive evaporation of fluid from the Sephadex layer.

dimensional separations, a single sample was spotted at the center of the above-mentioned imaginary line.

After the deposition of the sample the container was covered with a glass bell to prevent evaporation during the run. The time required for the development of a chromatogram was about 10–20 min/cm of length. At the end of this period, the flow of fluid was stopped by removing the filter paper pads indicated with number 3 and 6 in Fig. 2, the container was placed in a horizontal position, and the substances present on the chromatogram were detected by one of the techniques described below.

Zone electrophoresis

After the chromatographic separation, the container was placed between two troughs containing the buffer and the electrodes as indicated in Fig. 3. The container

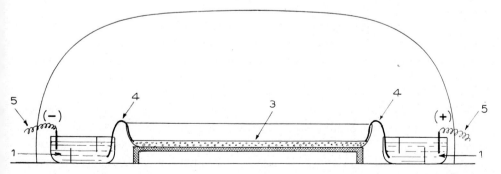

Fig. 3. (1) Troughs containing the electrodes and buffer. (2) Container supporting the Sephadex layer. The Sephadex layer must be at least 1 cm above the level of the buffer in the troughs. (3) Sephadex layer; (4) Filter paper soaked in buffer and disposed between the Sephadex layer and the buffer surrounding the electrodes; (5) Electrodes; (6) Protection bell.

References p. 211

was oriented so that the electric field lines were normal to the direction of the flow of the solvent during the previous chromatographic run. The ionic strength of the buffer used to prepare the Sephadex layers for zone electrophoresis ranged between 0.02 and 0.06 μ. The electrophoretic run was carried out with a potential gradient of 10–20 V/cm length of the Sephadex layer for 3–10 h, depending upon the mobility of the substances under investigation.

Detection of the separated substances

Upon completion of the chromatographic or electrophoretic run, a damp sheet of Whatman No. 1 paper was deposited on the Sephadex layer so as to obtain a perfect adherence between paper and gel. Under these conditions some of the chromatographed material passed from the Sephadex gel to the paper. After 40–50 min, the paper was carefully removed from the layer, dried and treated with Amido Schwarz B 10 according to WUNDERLY[7] for the detection of proteins, or with ninhydrin according to INGRAM[4] for the detection of peptides.

Controls with hemoglobin showed that about 20 % of the material present in the Sephadex layer is transferred to the paper.

Recovery of substances from the Sephadex layer

The gel corresponding to the zones of the layer which, according to the detecting paper, have been shown to contain the separated substances, were removed from the container with a spatula, placed in a small centrifuge tube and suspended in 5 ml of water or other suitable solvent. The test tubes were then centrifuged. Most of the analysed substance is found in the supernatant. If a more complete recovery is required the procedure can be repeated.

3. RESULTS AND DISCUSSION

Separation of peptides

A typical separation of the peptides obtained by tryptic digestion of hemoglobin is shown in Fig. 4.

As shown, chromatography by molecular filtration on a thin layer, associated with electrophoresis, makes it possible to obtain good separations of peptides.

It should be pointed out that with the present technique the chromatographic and electrophoretic runs can be carried out with the same buffer, so that it is not necessary to remove the solvent or buffer between one run and the next.

The method proposed for the detection of the spots on the chromatogram, moreover, uses only one fraction of the separated material (*i.e.* the fraction that passes from the gel to the detecting paper) so that a considerable portion of the separated substances can be eluted unaltered from the gel and used for further studies. This is particularly useful when, in the course of a study on the primary composition of a protein, it is desirable to study the amino-acid composition of the peptides obtained by partial hydrolysis.

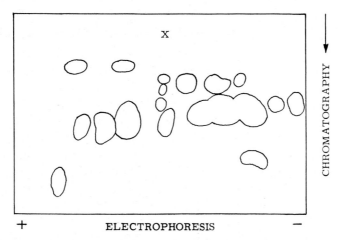

Fig. 4. Peptide analysis. Electrochromatogram on a thin layer of Sephadex G 25, fine, equilibrated with 0.02 M phosphate buffer pH 6.8; analyzed material: 0.1 ml of a 6% tryptic digest of human hemoglobin. Dimensions of the layer: 30 × 20 × 0.1 cm. Duration of the chromatographic run: 2 h 30 min. Electrophoretic run: potential = 460 V; intensity = 25 mA; temperature = 5° ±1 °C; duration = 7 h. The detecting paper was sprayed with ninhydrin. X = start.

The possibility of recovering a considerable amount of the separated material can also be of advantage when bidimensional techniques prove insufficient for achieving a complete separation. Under these conditions, in fact, it is possible to elute from the Sephadex the incompletely separated material and to submit it to further fractionation processes.

In this connection it should be remembered that the proposed method exploits molecular properties that are different from those generally used for fractionating peptides in studies of the peptide composition of proteins, and can therefore be a useful supplement to classical techniques based on partition chromatography.

Another field in which the recovery of the material after separation is of considerable interest is the study of the distribution of activities in enzymatic or hormonal protein mixtures. The catalytic activity of the different fractions can in fact be assayed on the material eluted unaltered from the gel.

Chromatography of proteins

A typical chromatogram is reproduced in Fig. 5 which shows that thin-layer chromatography on Sephadex G 100 permits the separation of proteins of molecular weight ranging from 25,000 to more than 2 million.

Thin-layer chromatography on Sephadex G 100 or G 200 also provides a method for the quick, though rough, evaluation of the molecular weight of proteins. A linear relation between the chromatographic mobility on thin layers of Sephadex G 25 and the molecular weight of amino acids and small peptides has been described by DETERMANN[2]. A similar relation has been found by us for proteins of molecular weight ranging from 25,000 to 140,000 using thin layers of Sephadex G 100 and G 200 (Fig. 6). The unknown substance is run on a Sephadex plate with several other proteins of

References p. 211

known molecular weight. It is then possible to draw a curve analogous to the one reported in Fig. 6 and to ascribe a molecular weight to the unknown protein by interpolation. This method of assessing molecular weight is considerably faster and less expensive than those based on column chromatography on Sephadex.

It should be noted that several factors can affect the behaviour of proteins during chromatography on thin layers of Sephadex. Among other factors, denaturation, electrostatic interaction between polar groups, and adsorption on the gel should be taken into consideration. The occurrence of denaturation can be detected by eluting the proteins from the Sephadex layer after the run and investigating their properties. Electrostatic interaction and adsorption can be evaluated by repeating the chromatographic separation using buffers of varying ionic strength or different solvents.

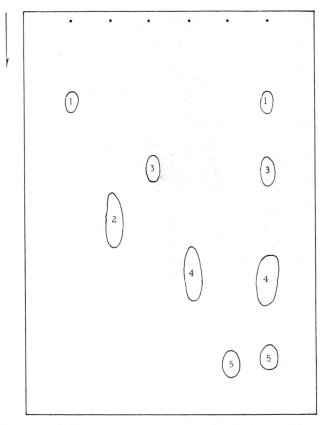

Fig. 5. Protein analysis. Chromatography of proteins on a thin layer of Sephadex G 100, equilibrated with 0.05 M phosphate buffer, pH 7.4. The analysis was carried out on 0.02 ml of each solution. Protein solutions were 0.5 %, the pyridoxamine solution was 0.1 %. (1) pyridoxamine, used as low molecular weight reference substance (completely diffusible within the dextran gel); (2) adult human hemoglobin (molecular weight = 68,000); (3) chymotrypsin (molecular weight = 25,000); (4) bovine serum albumin (dimer, molecular weight = 140,000, see ref. 8); (5) hemocyanin (molecular weight > 2,000,000) used as high molecular weight reference substance (completely excluded from the dextran gel). Dimensions of the layer: 30 × 40 × 0.1 cm. Duration of the run = 7 h. Temperature = 5° ± 1°C. The detecting paper was examined in U.V. light to detect pyridoxamine and then treated with Amido Schwarz B 10.

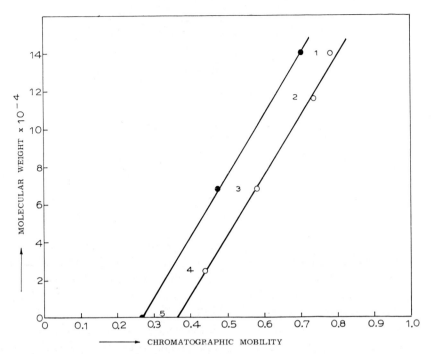

Fig. 6. Relation between molecular weight and chromatographic mobility on Sephadex G 100 and G 200. The chromatographic mobility is expressed as the ratio between the displacement of the substance and the displacement of hemocyanin (high molecular weight substance completely excluded from the gel). Mobility on Sephadex G 100 = ○ ○ ○; mobility on Sephadex G 200 = ● ● ● . The gel was equilibrated with 0.05 M phosphate buffer pH 7.4. (1) Albumin dimer (ref. 8); (2) aspartic-aminotransferase; (3) adult human hemoglobin; (4) chymotrypsin; (5) pyridoxamine.

This work is part of a program supported by "Impresa di Enzimologia del Consiglio Nazionale delle Ricerche".

REFERENCES

1. P. FLODIN, *Dextran Gels and their Applications in Gel Filtration*, Pharmacia, Uppsala, 1962.
2. H. DETERMANN, *Experientia*, 15 (1962) 430.
3. B. G. JOHANSSON AND L. RYMO, *Acta Chem. Scand.*, 16 (1962) 2067.
4. V. M. INGRAM, *Biochim. Biophys. Acta*, 28 (1958) 539.
5. H. LIS, *Biochim. Biophys. Acta*, 28 (1958) 191.
6. A. ROSSI FANELLI, E. ANTONINI AND A. CAPUTO, *J. Biol. Chem.*, 236 (1961) 397.
7. C. WUNDERLY, in *A Laboratory Manual of Analytical Methods in Protein Chemistry*, edited by P. ALEXANDER AND R. J. BLOCK, Vol. 2, p. 226, Pergamon Press, Oxford, 1960.
8. K. O. PEDERSEN, *Arch. Biochem. Biophys.*, suppl. I., (1962) 157.

THIN-LAYER CHROMATOGRAPHY ON SILICA GEL OF FOOD COLOURS

M. AMALIA CIASCA AND C. G. CASINOVI

Istituto Superiore di Sanità, Department of Biological Chemistry, Rome (Italy)

1. INTRODUCTION

A new list of dyes permissible as food additives in foodstuffs was published early this year by the Italian Government, modifying the already existing list and including a number of dyestuffs whose use was permitted in other European countries[1].

Since the colouring matter added to foodstuffs must, according to the new law, be declared, easier and more rapid control is possible.

It is known that food colours may be separated and identified by means of paper chromatography[2]. More recently WOLLENWEBER reported their separation by thin-layer chromatography on cellulose[3] to be an excellent and rapid method.

It is our opinion, however, that in the field of food analysis more than one analytical method should be available. For this reason we have recently examined the possibilities of applying thin-layer chromatography on silica gel for food colours, in the hope of establishing a method to be used where the more conventional techniques quoted above would be impracticable. For instance, the differences of adsorption power of cellulose and silica gel may play an important role when accompanying impurities must be separated.

So far the method has been tested with fifteen dyes (see Table 2).

2. METHOD AND RESULTS

Silica gel layers were prepared according to STAHL[4] and several solvent systems were tried in order to establish the best conditions for separation. Results are reported in Table 1.

As can be seen from Table 1 only the last two solvents (solvents 1 and 2) gave satisfactory results. It is also clear that some of the dyestuffs examined are contaminated to a greater or lesser degree by secondary components. The R_F values for the substances tested are reported in Table 2.

From the results reported above, it is seen that identification of the dyes examined is quite easy when dealing with single compounds or simple mixtures. Of course, in some particular cases (*e.g.* mixtures of the three yellows: chrysoidine, quinoline yellow, fast yellow) characterisation of all the components may offer some difficulties.

The suitability of the method for analysing extracts of foodstuffs will be investigated as well as comparative tests with other dyes (of natural origin) present in the Italian list of permitted food colours.

TABLE 1

EXAMINATION OF SOLVENT SYSTEMS FOR SEPARATION OF FOOD COLOURS

sodium citrate (2.5 %) : ammonia (25 %) = 4 : 1	Too high R_F values
n-butanol : water	Too low R_F values
n-butanol : acetic acid : water = 4 : 1 : 5	Too high R_F values
amyl alcohol : acetic acid : water = 3 : 2 : 2	Too low R_F values, extensive trailing
n-butanol : ethanol : pyridine : water = 12 : 12 : 1 : 4	Extensive trailing
n-butanol : ethanol : diethylamine : water = 12 : 12 : 1 : 4	Extensive trailing
isobutanol : ethanol : ammonia (25 %) : water = 12 : 12 : 4 : 4	Extensive trailing
n-butanol : ethanol : ammonia (25 %) = 20 : 20 : 5	Solvent demixing on the plate
ethanol : ammonia (25 %) = 40 : 1	Too high R_F values
butanol : ammonia (25 %) : water = 25 : 4 : 1	Too low R_F values
n-butanol : ethanol : water = 20 : 20 : 5	Satisfactory resolving power, solvent 1
n-butanol : ethanol : water : HCl (0.3 N) = 20 : 20 : 5 : 1	Satisfactory resolving power, solvent 2

TABLE 2

R_F VALUES OF FOOD COLOURS IN TWO SOLVENTS

Solvent 1 = n-butanol : ethanol : water = 20 : 20 : 5.
Solvent 2 = n-butanol : ethanol : water : HCl(0.3 N) = 20 : 20 : 5 : 1.

dyes	Schultz[5]	C.I.[6]	R_F	
			in solvent 1	in solvent 2
Tartrazine	737	19140	0.13–0.33 (trail)	0.04
Chrysoidine	186	14270	0.65	0.64
Quinoline yellow	918	47005	0.68	0.56
Fast yellow	172	13015	0.51	0.46
Orange S	—	15985	0.51 (trail)	0.0–0.28 (trail)
Orange GGN	—	15980	0.51	0.28
Azorubine red No. 2	208	14720		0.24–0.36
Amaranth	212	16185	0.17	0.0
Ponceau 4 R	213	16255	0.37	0.12
Scarlet GN	—	14815	0.55	0.0
Ponceau 6 R	215	16290	0.0	0.0
Indantrene blue RS	1228	69800	0.0	0.0
Patent blue V	826	42045	0.32	0.28
Indigo carmine	1309	73015	0.0	0.0–0.03
Brilliant black	—	28440	0.0 (trail)	0.0 (trail)

3. SUMMARY

Thin-layer chromatography on silica gel has been applied to the separation and the identification of food colours from the new list of permissible colours for foodstuffs of the Italian Government.

REFERENCES

1. D.M. of 19.1.1963, Gazz. Uff., 7th March, 1963.
2. *Separation and Identification of Food Colours*, Ass. of Public Analysts, 1960.
3. P. WOLLENWEBER, this Volume p. 14.
4. E. STAHL, *Angew. Chem.*, 73 (1961) 646.
5. G. SCHULTZ, *Farbstofftabellen*, Leipzig, 1931.
6. Colour Index, Bradford, 1956.

THE APPLICATION OF THIN-LAYER CHROMATOGRAPHY TO INVESTIGATIONS OF ANTIFERMENTATIVES IN FOODSTUFFS

M. COVELLO AND O. SCHETTINO

Istituto di Chimica Farmaceutica e Tossicologica dell'Università di Napoli (Italia)

PRELIMINARY NOTE

1. INTRODUCTION

In the framework of the struggle against adulteration of foodstuffs, research on the antifermentatives, a category which includes substances which are chemically very diverse, occupies a significant position. This may primarily be the result of the very rapid and broad development of the food-preserving industry.

Legislation in all countries provides more or less drastic limitations concerning the uses of antifermentatives which vary considerably from country to country, and this in itself renders the research undertaken to reveal possible frauds more complex. Without entering into the legal aspects of the problem we may observe that, even for substances whose use is permitted within certain determined limits, one may impose an identity control which will have to be followed by a quantitative determination.

In fact the analyst's main difficulty is due to the large number of substances that can be employed for preserving purposes. Indeed, the need for establishing the identity of a chemical compound through numerous tests has been proposed; each test would then deal specifically with one of the substances capable of being used for antifermentative action. This makes it essential to examine a great number of samples of the product, and therefore involves considerable expenditure of both reagents and time.

Further complications arise in the case of the presence of more than one antifermentative: some commercial preservatives are made up from a mixture of chemical compounds (*e.g.* the well-known Nipakombin, a mixture of esters of *p*-hydroxy benzoic acid) which are difficult to distinguish from one another. Moreover, the coupling of preservatives may be undertaken with the aim of exploiting a particular form of synergism, which, as demonstrated in quite recent work[1], allows both broadening of the antimicrobial spectrum of antifermentatives commonly used and the reduction of the absolute amounts used.

The sum of these difficulties, which are of course largely of a practical nature, has caused many authors to suggest procedures involving only a single series of operations by which it may be possible to identify groups of antifermentative substances. A combination of traditional chemical, and paper chromatographic methods has been suggested.

The recent development of thin-layer chromatography, has induced us to make a contribution to the solution of the problem. The method is particularly suitable for such studies since it is rapid, simple, and amenable to special techniques which are not practicable on paper.

TABLE 1

PHYSICAL PROPERTIES OF ANTIFERMENTATIVES WHICH MAY BE SUBLIMED

Antifermentative	Melting point °C	Sublimation temperature °C	λ_{max} in ethyl alcohol mμ
(1) Sorbic acid	134.5	55	248
(2) Dehydroacetic acid	109–111	60	231 and 295
(3) Benzoic acid	122.4	60	225
(4) p-Hydroxybenzoic acid	213	80	259
(5) p-Chlorobenzoic acid	243	80	232
(6) Salicylic acid	159	76	230 and 298
(7) Cinnamic acid	133–136	90	265
(8) Methyl p-hydroxybenzoate	125–128	70	259
(9) Propyl p-hydroxybenzoate	95–98	70	259

We have studied a group of substances currently used as antifermentatives which have the common property of subliming in the range 55°–90° C (see Table 1). Some investigators[2-5] have used this property for purposes of identification. Except for an original spectrophotometric method proposed by STANLEY[6] (and continued by JAULMES and others[7]) for benzoic acid, which is based on the differential measurement of its striking characteristic properties, the recent work is notable for the combination of paper chromatography with at least one other method of study[8-10].

In the present work, the substitution of the thin layer for the filter-paper technique is not to improve the chromatographic separation of the compounds examined, which is the same in both procedures, but is rather to facilitate the thermal treatment (sublimation). Details will be found in the experimental section. By this method it was possible to recover the antifermentatives in the pure state, thus facilitating unequivocal identification.

It should be pointed out that the official Italian analytical standards for food products do not at the moment admit chromatographic results as valid evidence of identification of antifermentatives, although they do recognise a sublimation process like those suggested by the authors cited above.

The method we propose would permit the analyst to express the results of his analyses carried out using chromatography and still satisfy the official requirements.

2. EXPERIMENTAL

The antifermentatives are extracted from the foodstuffs, chromatographed on a thin layer of silica gel and removed after separation by sublimation according to BAEHLER's technique[11]. A copy of the chromatogram is thus obtained, and the position of the substance sought is located by observation of the sublimed crystals.

Subsequent microscopic, spectrophotometric or chemical identification is rendered easy and is not prejudiced by the previous use of chemical detectors. Moreover, it is possible to carry out a quantitative determination on a second chromatogram, run simultaneously, by means of elution and spectrophotometry.

Extraction of the antifermentative

The classic procedures can be used for these substances without any substantial modification. The foodstuff is acidified and extracted with equal volumes of ethyl ether and petroleum ether. The extract is subsequently purified by washing with aqueous alkali followed by acidification of the washing water and re-extraction with ether.

The ether extract, reduced to a small volume by spontaneous evaporation in the air or by a forced air current can be applied to the chromatographic plate.

The size of sample to be taken may be decided after consideration of the kind of product under examination (liquid, solid, paste) and of the expected content of the antifermentative. The flexibility of the chromatographic technique employed is emphasised for it permits quantities of preservative substances ranging from ten to two or three hundred μg to be used without substantial loss of sensitivity in definition.

Chromatography

The equipment made by C. Desaga Gmbh. of Heidelberg was used. STAHL[12,13] considers it suitable for thin-layer chromatography. This equipment allows the preparation of absorbent layers of thicknesses varying from 250 to 2000 μ.

In all our experiments we used layers of Merck silica gel G containing calcium sulphate as binder. Metal plates (strongly chromed brass) measuring 5×20 cm with a thickness of 5 mm proved to be more suitable for carrying out the thermal treatment of the successive phases without in any way interfering with the chromatographic process. On these plates a layer of 250 μ of the paste (obtained by shaking a mixture of 25 g of silica gel G and 50 ml of distilled water) was spread by means of the Desaga manual stratificator. This layer was left to dry in air and completed in a vacuum drier; it was not considered necessary to submit it to activation by heating.

The ether extract (as mentioned above), is placed by means of a capillary pipette 2 cm from the base of the plate and, after complete elimination of the solvent, the development is commenced. In general monodimensional chromatography is sufficiently selective; in doubtful cases, however, one can use bidimensional development, by employing 20×20 cm plates.

References p. 219

The solvents used for the development were as follows:
(1) Butanol saturated with 2 N ammonia
(2) Isopropyl alcohol–ethyl alcohol–conc. ammonia–water (5 : 2 : 0.5 : 1)
(3) Benzene–acetic acid–water (4 : 9 : 2).

The R_F of the antifermentatives studied are given in Table 2. One may observe that the best separations are obtained with solvent no. 1. Solvent no. 3, because of its acidity, is not suitable for chromatography on metal plates which slowly become corroded.

The reproducibility of the results is assured by the complete saturation of the atmosphere in the developing tank; this is achieved by means of strips of filter paper fixed to the internal walls and floating in the developing liquid. Under these conditions the average development time for a length of 10 cm is about three hours.

TABLE 2

R_F-VALUES OF SUBLIMABLE ANTIFERMENTATIVES

Solvent 1: Butanol saturated with 2 N ammonia
Solvent 2: Isopropyl alcohol–ethyl alcohol–conc. ammonia–water (5 : 2 : 0.5 : 1)
Solvent 3: Benzene–acetic acid–water (4 : 9 : 2)

Antifermentative	R_F		
	Solvent 1	Solvent 2	Solvent 3
(1) Sorbic acid	0.23	0.77	0.74
(2) Dehydroacetic acid	0.27	0.81	0.68
(3) Benzoic acid	0.24	0.65	0.88
(4) p-Hydroxybenzoic acid	0.12	0.75	0.74
(5) p-Chlorobenzoic acid	0.24	1	1
(6) Salicylic acid	0.53	0.79	0.70
(7) Cinnamic acid	0.43	0.72	0.61
(8) Methyl p-hydroxybenzoate	0.66	0.88	0.72
(9) Propyl p-hydroxybenzoate	0.67	0.90	0.71

Detection

BAEHLER's technique was adopted without substantial modification. This technique, employed by the author in the detection of purine compounds, involves heating the chromatographic plate to effect the sublimation of the chromatographed substance which is then condensed on a cold surface supported above the plate. When development was complete, the chromatoplate was carefully allowed to dry, then placed in a heating chamber consisting of a metal block heated by a gas flame and fitted with a thermometer well. A glass plate supported about 1 cm above the chromatographic plate by four diamond shaped discs served as condensation surface. This surface was cooled by setting a dish of glass or metal on its upper surface, and arranging a flow of cold water through this dish by means of a syphon or water pump.

By careful control of the heating block temperature, and checking (as suggested by BAEHLER) of the temperature of the upper surface of the chromatographic plate it is possible to achieve regular sublimation yielding crystalline deposits only slightly larger than the original spots. (The temperature at the surface of the chromatographic plate is usually about 60° lower than that of the heater, and may be determined by tests using materials of known melting point.)

R_F values may be calculated by marking on the condenser plate the starting point and solvent front of the original. Tests such as microscope examination of the crystal structure, microscopic melting point, spectrophotometric studies and specific chemical tests may all be carried out on the sublimed crystals.

Quantitative determination

Even when the sublimation is carried out with great care, the crystals obtained constitute only an aliquot of the material originally present. Thus quantitative determinations are not possible even though the crystals may easily be dissolved in appropriate solvents. For quantitative analysis it is thus necessary to run two separate chromatograms, one of which, treated as described above, will yield material for qualitative determination and precise location on the chromatogram. The second chromatogram may be eluted, and the antifermentatives determined spectrophotometrically. Quantitative aspects of the method described will be discussed in a further communication.

3. SUMMARY

A group of substances commonly used as antifermentatives has been studied. Such substances were extracted from foods containing them and chromatographed on a thin layer of silica gel. After separation they were isolated from the support by sublimation. Qualitative and quantitative identification by microscopy and spectrophotometry were facilitated and the use of reagents avoided.

REFERENCES

1. H. J. REHM, W. WITTMAN AND U. STAHL, *Z. Lebensm. Untersuch. Forsch.*, 115 (1961) 244.
2. R. FISCHER AND F. STAUDER, *Mikrochemie*, 8 (1930) 33.
3. R. CULTRERA, *Ind. Ital. Conserve*, 8 (1933) 298.
4. G. SERIS, *Ann. Fals. Fraudes*, 44 (1951) 373.
5. E. TRIFIRÒ, *Ind. Conserve Parma*, 35 (1960) 279.
6. R. L. STANLEY, *A.O.A.C. Journ.*, 43 (1960) 587.
7. P. JAULMES, R. MESTRES AND B. MAUDRON, *Ann. Fals. Expert. Chim.*, 54 (1961) 84.
8. J. L. JOUX, *Ann. Fals. Fraudes*, 50 (1957) 205.
9. I. P. GODIJN, *Z. Lebensm. Untersuch. Forsch.*, 115 (1961) 534.
10. A. S. KOVACS AND P. DENKER, *Ind. Obst-Gemuseverwert.*, 47 (1962) 1.
11. B. BAEHLER, *Helv. Chim. Acta*, 45 (1962) 309.
12. E. STAHL, *Chemiker Ztg.*, 82 (1958) 10.
13. E. STAHL, *Z. Anal. Chem.*, 181 (1961) 303.

DIREKTE QUANTITATIVE BESTIMMUNG VON KATIONEN MITTELS DÜNNSCHICHTCHROMATOGRAPHIE*

H. SEILER

Institut für Anorganische Chemie, Universität Basel (Schweiz)

ZUSAMMENFASSUNG

Es wurden vergleichende Versuche zur quantitativen Bestimmung von Cu, Co, Ni und Na, K, Mg auf Dünnschichtchromatogrammen mit verschiedenen Methoden durchgeführt:

1. durch Vergleich der Fleckengrössen,
2. durch Messung der Lichtabsorptionen der angefärbten Metallionen in durchfallenden Licht und
3. durch radiometrische Messung der entsprechenden radioaktiven Isotopen.

---

* Vollständiger Text dieser Arbeit: *Helv. Chim. Acta*, 46 (1963) 2629.

THIN-LAYER CHROMATOGRAPHY OF INORGANIC IONS

I. SEPARATION OF METAL DITHIZONATES

M. HRANISAVLJEVIĆ-JAKOVLJEVIĆ AND I. PEJKOVIĆ-TADIĆ

Faculty of Sciences, University of Belgrade (Yugoslavia)

AND

K. JAKOVLJEVIĆ

Institute of Hygiene, PRS (Yugoslavia)

1. INTRODUCTION

The thin-layer chromatography of inorganic ions has not been studied extensively. There are only few papers dealing with this problem, predominantly those from SEILER's group[1]. The cations of all analytical groups are reported to be separated on specially purified silica in rather polar solvent mixtures. The time required for these procedures varied from 20 min to 2 h. In addition to some tailing and sometimes small R_F value differences, the main disadvantages of these procedures were the relatively long duration of analysis and the time-consuming purification of silica gel.

For that reason we started a systematic study on thin-layer chromatography of inorganic ions in the form of their complex salts, primarily in the form of dithizonates. This paper deals with the separation of metal dithizonates by TLC. Diphenylthiocarbazone, usually abbreviated to "dithizone" produces brilliant yellow, red and violet coloured complexes with a dozen or more metals. It was used, with greater or less success, for the separation of some cations by column chromatography[2-5] and recently also by paper chromatography[6]. Taking dithizone, we had in mind that it is not a specific reagent and therefore could be very useful as an extracting reagent for a large group of elements. At the same time we hoped to avoid to a great extent the difficulties encountered in the previously mentioned procedures, namely, that the metal dithizonates, which have a very pronounced organic character, would be expected to behave rather like organic substances on chromatoplates.

2. DISCUSSION

Thin-layer chromatographic procedure applied to metal dithizonates gave very good results. The separation of mercury-, lead-, copper-, bismuth-, cadmium- and zinc-

References p.224

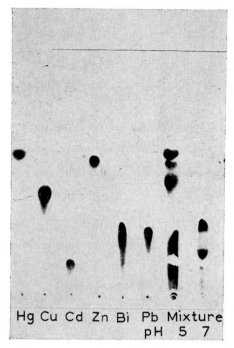

Fig. 1. Thin-layer chromatography of metal dithizonates. Conditions: chromatoplates (13 × 20) made of standard silica gel G according to Stahl; solvent system: benzene–methylene chloride (50 : 10, v/v); 10 μl of chloroform extracts of individual metal dithizonates as well of the mixture applied on the plate; running time: 40 min; solvent front: 13 cm. At pH 7 the spot of excess of dithizone can be seen.

dithizonate on chromatoplates was achieved within 20 to 40 min (see Fig. 1 and Table 1). The procedure involves two steps:

(a) formation and extraction of metal dithizonates and

(b) separation with simultaneous identification of individual cations.

Solutions of metal acetates or chlorides (0.1%) and a chloroform solution of dithizone (0.1%) were used in all experiments. Extractions were performed in acidic,

TABLE 1

R_F-VALUES OF METAL DITHIZONATES

Solvent mixture: benzene–methylene chloride (50 : 10)

Ion	Dithizone colour	$R_F \cdot 100$
Cd^{2+}	orange	13
Bi^{3+}	pink-orange brown	37
Pb^{2+}	red	34
Cu^{2+}	green-brown	48
Zn^{2+}	purple-red	50
Hg^{2+}	orange	58

and neutral media. Mercury, copper, zinc, cadmium and lead were extracted at pH 5, and bismuth at pH 7. However, some coextraction could not be avoided and some bismuth was extracted in acidic medium; on the other hand some lead could be found in the extract at pH 7. The main problem was imposed by lead because this ion was easily lost if the pH was not controlled accurately.

The chromatoplates were made of standard silica gel G according to STAHL and run in a benzene–methylene chloride solvent mixture. As metal dithizonates are differently and characteristically coloured, there was no need for any spraying reagent. The excess of dithizone did not interfere in the identification of cations and need not be removed. However, the removal of dithizone was easily achieved by shaking chloroform extracts with 0.02 N nitric acid or 0.02 N ammonium chloride solution.

Taking into account that this new procedure does not involve either the classical separation into analytical groups or any spraying reagent, we may say that the "dithizone method" represents an effective, simple and rapid technique for the separation and identification of a large number of elements.

The possibility of analysing other elements (such as Ag, Ni, Co etc.) by this method is under investigation and further results will be reported in due course.

3. EXPERIMENTAL

Materials

0.1% aqueous solution of metal acetates or chlorides; 0.1% chloroform solution of dithizone.

Solvent systems

benzene–methylene chloride (50 : 10 v/v) or
benzene–chloroform (50 : 5 v/v) or
benzene–carbon tetrachloride (50 : 5 v/v).

Running time

20–40 min depending on the cations present in the mixture.

Procedure

Equal volumes of salt solutions acidified with glacial acetic acid to pH 5 and a chloroform solution of dithizone were shaken vigorously for about three minutes. The separated chloroform extract of mercury-, lead-, copper-, cadmium- and zinc-dithizonate was washed with 0.02 N nitric acid to remove the excess of dithizone*. The subsequent extraction was carried out from the solution which was previously neutralized by addition of dilute ammonia (1 : 20) to pH 7. This extract was freed from dithizone by means of 0.02 N ammonium chloride solution. The extractions in the same medium were repeated several times *i.e.* to exhaustion.

---

* Avoid excessive washing when lead is present.

References p. 224

Thin-layer chromatography was carried out on glass plates (13 × 20) coated with a 0.2-mm thick layer of silica gel G according to STAHL. On the air dried chromatoplates, 10 µl of chloroform extracts were applied along a line 2 cm from the lower edge of the plate and 1.5 cm apart from each other. The development was carried out in previously equilibrated chromatographic tanks (24 × 16 × 8) and complete separation was achieved in 20–40 min.

4. SUMMARY

Thin-layer chromatography for the separation and identification of inorganic cations is described. Mercury-, lead-, copper-, bismuth-, cadmium- and zinc-dithizonate were separated on standard silica gel layers in a benzene–methylene chloride solvent system. Detection called for no spraying reagent since metal dithizonates have different and characteristic colours.

REFERENCES

1. E. STAHL, *Dünnschicht-Chromatographie*, Springer, Berlin, 1962, p. 481–496.
2. O. ERÄMETSÄ, *Suomen Kemistilehti*, 16B (1943) 13; *C.A.*, 40 (1946) 4620-8.
3. M. TANAKA, T. ASHIZAVA AND M. SHIBATA, *Chem. Researches (Japan), Inorg. and Anal. Chem.*, 5 (1949) 35; *C.A.*, 43 (1949) 8945h.
4. TAKASHI ASHIZAWA, *Bunseki Kagaku*, 10 (1961) 851; *C.A.*, 56 (1962) 9399a.
5. A. P. SEYFANG AND J. E. DUNABIN, U.K. At. Energy Authority, Ind. group SCS–R–134, 1959; *C.A.*, 54 (1960) 11632d.
6. HIDEO NAGAI, *Kumamoto J. Sci.*, Ser. A4 (1960) 256; *C.A.*, 54 (1960) 24089b.
   G. VENTURELLO AND A. M. GHE, *Anal. Chim. Acta*, 10 (1954) 335; *C.A.*, 48 (1954) 8112a.

INDEX

Absorbent zones, device for collecting, 39
Accelerated chromatography, 93
Acenaphthene, 136
Acenaphthylene, 133, 136
Acetaldehyde, 138, 139
— 2,4-dinitrophenylhydrazone, 142
Acetone, 30, 138, 139
— 2,4-dinitrophenylhydrazone, 142
Acetylated cellulose powder, 23
cis-Aconitate, 25
Aconitine, 150
Acraldehyde, 138, 139
— 2,4-dinitrophenylhydrazone, 142
Acridine, 135
Acylneuraminic acids, 30
Adenine, 28
Adenosine, 28
ADP, 29
Adsorption and crystallization of substances, 126
Adsorptivity, 41
Agar, 27
Alanine, 167
β-Alanine, 167
—, TLE, 63
Aldehydes, 109
Aldosterone, 86, 183
Aliphatic amines, 20, 21
Alizarin S, TLE, 65
Alkaloids, 8, 20, 40, 91, 155
Alkanals, 109
Alkanones, 109
Alkoxy diglycerides of dogfish liver oil, 104
Alkyl hydroperoxides, 109
Alumina, 90
—, activity determination, 90
—, activity titration, 37
Aluminium oxide, 10, 14
Amaranth, 213
Amines, 94
—, TLE, 62, 64
Amino acids, 20–23, 30, 91, 165
—, basic, 171
—, TLE, 62–64
Aminoaciduria investigation, 165
p-Aminoazobenzene, 38
α-Aminobutyric acid, 167
β-Aminobutyric acid, 167
γ-Aminobutyric acid, 167
α-Aminoisobutyric acid, 167
β-Aminoisobutyric acid, 167, 173
γ-Aminoisobutyric acid, 173
4-Amino-4'-nitroazobenzene, 79
2'-AMP, 29
3'-AMP, 29
5'-AMP, 29
n-Amylaldehyde, 138

Analgesics, 40
Androstane-3,17-dione, 183
Androst-4-ene-3,17-dione, 78
Androsterone, 183
p-Anisaldoximes, 160
Anisic acid, TLE, 61
Anthocyanine, 30, 92
Anthracene, 133
Anthraquinone dyes, 23
Antifermentatives, 215, 218
Antioxidants, 92, 108, 109
Apparatus for centrifugal TLC, 119
— for circular chromatography, 122 ff.
— for TLE, 56
Application of substance, 124
Applications of TLC, 1
— —, review, 11
Applicator for loose layers, 33
Applicators, 88
Arabinose, 25 ff.
L-Arabinose, 26
Arginine, 167
Artemisia absinthium fatty acids, 103
— — fatty esters, 103
Asparagine, 167
Aspartate, 24, 25
Aspartic acid, 167
Aspartic-aminotransferase, 211
Aspergillus flavus, toxic material, 110
ATP, 29
Azobenzene, 38
Azo dyes, 38, 40
—, lipophilic, 79
Azorubine red No. 2, 213

Barbiturates, 40
Basic fractions from plants, 158
Belladonna alkaloids, 21, 40
7,8-Benzoquinoline, 135
Benzoic acid, 218
B.H.T., 109
Bile acids, 78, 96
Biotine, 109
Bismuth, 222
B.N.A., 109
BN-chamber, 7
Borate, impregnation of layers, 90
Boric acid, impregnation of layers, 90
Brassicasterol, 198
Brilliant black, 213
Brominated triglycerides, 91
Bromophenol blue, TLE, 65
Brucine, 150, 156
Butter yellow, 119
n-Butylamine, 21
—, TLE, 63

INDEX

n-Butyraldehyde, 138
— 2,4-dinitrophenylhydrazone, 142

Cadaverine, TLE, 64
Cadmium, 222
Caffeic acid, TLE, 59, 60
Calcium carbonate, 92
— hydroxide, 92
— pantothenate, 109
Caproic aldehyde, 138
Caprylaldehyde, 139
— 2,4-dinitrophenylhydrazone, 142
Carbamide, 10
Carbazole, 134
Carbohydrates, polycyclic, 20, 29
—, polynuclear aromatic, 132
Carbonyl compounds, 118
— —, volatile, 40
Carob bean flour, see Johannisbrotkernmehl
α-Carotene, 109
β-Carotene, 40, 109
Carotenoids, 90, 109
— aldehydes, 109
Carotol, 40
Carrageenin, 27
(+)-Catechin, 27
Catharantus roseus alkaloids, 40
Cations, 220
Cellulose, 79
—, carboxymethyl, 15
—, DEAE, 15, 92
—, diethylaminoethyl, 15, 92
—, ECTEOLA, 15
—, exchangers 15
—, layers, 14
—, modified, 92
—, partially acetylated, 91
—, phosphorylated, 15
—, powder, 23, 91
— —, highly acetylated, 91
— —, MN, 10, 15
Centrifugal TLC, 119
Cephalins, 108
Cerebrosides, 108
Chelidonium alkaloids, 21
Chlorella, C_{16} fatty acids from, 101
—, C_{18} fatty acids from, 101
p-Chlorobenzoic acid, 218
Chlorogenic acid, TLE, 61
12-Chloro-13-hydroxy-9-octadecanoic acids, 103
13-Chloro-12-hydroxy-9-octadecanoic acids 103
threo-12,13-Chlorohydroxyoleic acid, 103
threo-13,12-Chlorohydroxyoleic acid, 103
6-Chloropurine, 28
Chocolate extracts, 27
Cholestane, 78
Cholestanol, 53
Cholesterol, 42, 78, 79, 100, 110, 198, 202
— acetate, 78
— esters, 110
Cholesteryl oleate, 100
Chromatographic data, standardised, 116

Chromotographic number, 116
Chromatography by molecular filtration, 206
Chromatoplates, 8
—, development of, 92
—, preparation of, 88
Chromatostrip, 8, 87
— technique, 3
Chromium complexes, 40
Chromotropic acid, TLE, 62
Chrysene, 133
Chrysoidine, 213
Chymotrypsin, 211
Cinchonidine, 150
Cinchonine, 150
Cinnamic acid, 92, 218
—, TLE, 60
Circular technique, 122 ff.
— —, application of, 125
Citrate, 25
Citrulline, 167
3'-CMP, 29
Cobalt, 220
— complexes, 40
Cocaine, 150
Cocoa butter, 91, 105
— substitute, 105
Codeine, 150
Colchicine, 150
Colour reactions, 82
α-Colubrine, 156
β-Colubrine, 156
Comparison of thin layers for TLE, 55
Condensed phosphate, 31
Congo red, TLE, 65
Continuous elution, 93
Co-ordination, 42
Copper, 220, 222
Coprostanol, 53
Corrosive reagents, 95
Cortexone, 78, 183
Corticosteroids, 78, 85, 180, 183
—, mixtures, quantitative analysis, 190
—, quantitative determination, 188
Corticosterone, 183
Cortisol, 78
Cortisone, 78, 183
Cresols, 40
Cryptoacoronol, 40
Crystal ponceau, TLE, 65
p-Cumaric acid, TLE, 59, 60
p-Cuminaldoximes, 160
Cyclopropene fatty acids, 105
Cysteic acid, 167
Cysteine, TLE, 64
Cystine, 167
—, TLE, 63

DEAE-Sephadex, 10, 92
Dehydroacetic acid, 218
Dehydrobrucine, 156
Dehydrocholesterol, 202
7-Dehydrocholesterol, 198, 202
Dehydroepiandrosterone, 183
Dehydroskythantine, 156

Detection, 95
— methods, 80
— reagents for steroids, 80
Development of chromatograms, 124
— of TLC, 1
Dexamethasone, 183
Dextrans, 91
Diaboline, 156
Diacyl peroxides, 109
Dialkyl peroxides, 109
2′,7′-Dibromofluorescein, 95
Dicarboxylic acids, 92
2,7-Dichlorofluorescein, 98
2′,7′-Dichlorofluorescein, 95
cis-9,10-cis-12,13-Diepoxystearic acid, 103
α,α'-Diethyl-4,4′-stilbenediol, 195
Diethyl stilboestrol, 195
Diffusion between chromatoplates, 49
Diglycerides, 97
1,3-Diglycerides, 104
Diglycerides, alkoxy, of dogfish liver oil, 104
—, isomers, 104
Dihydrocholesterol, 198
Dihydro-β-sitosterol, 198
Dihydrosphingosine, 108
Dihydroxyacetone, 139
— 2,4-dinitrophenylhydrazone, 142
Dihydroxy fatty acids, *threo* and *erythro* isomers, 103
1,3-Dihydroxynaphthalene, TLE, 62
$3\alpha,21$-Dihydroxypregnane-11,20-dione, 183
$17\alpha,21$-Dihydroxypregnane-3,20-dione, 183
$17\alpha,21$-Dihydroxypregnane-3,11,20-trione, 183
$17\alpha,21$-Dihydroxy-Δ^4-pregnene-3,20-dione, 183
3β-Dimethylaminocholest-5-ene, 42
2,6-Dimethylnaphthalene, 133
Dimethyl yellow, TLE, 65
2,4-Dinitrophenylhydrazones, 118, 119, 138
Diolein, 100
Dioleoyl lecithin, 100
o-Diphenols, TLE, 59
Diphenyl, 133
Diphenylene oxide, 134
— sulphide, 134
Dipoles, 42
Double bonds, isomerisation on alumina, 90
Dyes, 20, 23, 94, 119
—, TLE, 64, 65
—, fat soluble, 40

Elaidodistearin, 106
Eleodistearin, 106
Eluotropic graduation, 78
— series, 41, 75, 92, 93
(—)- Epicatechin, 27
Epicholestanol, 53
Episulphido fatty acids, 102
Epoxy acids, isomers, 102
Epoxydocosanoic acids, isomers, 103
9,10-Epoxy-9-octadecanoic acid, 103
12,13-Epoxy-9-octadecanoic acid, 103
cis-12,13-Epoxyoleic acid, 103

cis-9,10-Epoxystearic acid, 103
Equieluotropic series, 77
— solvent system, 78
Ergosterol, 198, 202
Ergot alkaloids, 20, 21
Ergotamine, 21
Ergotoxin group, 21
Eriochrome black T, TLE, 65
Essential oils, 87
Esterases, 94
Ester linkage hydrolysis on alumina, 90
Esters, 96
Ethanolamine, 21
—, TLE, 63, 64
Ethoxystrychnine, 156
Ethylamine, 21
—, TLE, 63, 64
Ethylenediamine, TLE, 63
Ethyl gallate, TLE, 60
Ethylmorphine, 150
Ethyl protocatechuate, TLE, 60
Etiocholane-$3\alpha,11\beta$-dihydroxy-17-one, 183
Etiocholane-3α-hydroxy-11,17-dione, 183
Etiocholanolone, 183
Experimental technique, 4

Factice, 92
Fast yellow, 213
Fats, minor constituents, 110
Fatty acids, 92, 100
— —, acetylenic, 105
— —, in butter, 99
— —, C_{16} fraction from *Chlorella*, 101
— —, C_{18} fraction from *Chlorella*, 101
— —, cyclopropene, 105
— —, epoxy, 105
— —, halohydroxy, 103,
— —, hydroxy, 105
— —, α-hydroxy, 103
— —, keto, 105
— —, methyl esters, 91, 102
— —, — —, from *Chlorella*, 101
— —, — —, critical pairs, 101
— —, — —, degree of unsaturation, 102
— —, normal, 102
— —, polyhydroxy, 103
— —, unsaturated, autoxidation of methyl esters, 99
— —, —, chlorohydroxy, 102
— —, —, mercuric acetate adducts, 99, 101
— —, —, methyl esters, 90
— —, —, stereoisomers of hydroxy, 102
Fatty esters, α-hydroxy, 103
FC 28 F phytosterols, 202
Ferulic acid, TLE, 61
Filix phloroglucidene, 122
Flavonoids, 96
Fluoranthene, 133, 134
Fluorene, 134
Fluorescein, 82
—, TLE, 65
Fluorescent layers, 82
Fluoroacetate, 25
Foods, 110

Food-colouring agents, 22, 110, 212
— —, fat soluble, 117
Formaldehyde, 138, 139
— 2,4-dinitrophenylhydrazone, 142
Formamide, impregnation of layers with, 91
Fructose, 27
Fumarate, 24, 25

Galactose, 25–27
D-Galactose, 26
Galacturonic acid, 25 27
Gallates, 109
Gallic acid, 109
— —, TLE, 60
Gangliosides, 31, 108
Gels, 20, 26
—, hydrolysed, 26, 27
Gentisic acid, TLE, 59,60
Germacrane derivatives, 40
Glucose, 25, 27
D-Glucose, 26
Glucuronic acid, 25–27
Glutamate, 24, 25
Glutamic acid, 167
— —, TLE, 64
Glutamine, 167
Glyceraldehyde 2,4-dinitrophenylhydrazone, 142
Glycerides, 92
—, brominated, 95
—, cocoa butter, 106
—, corn oil, 106
—, hydrogenated, 95
—, lard, 106
—, olive oil, 106
—, partial, 97, 104
—, soya bean oil, 106
Glycine, 167
—, TLE, 63,64
Glycolaldehyde, 138, 139
— 2,4-dinitrophenylhydrazone, 142
Glycollic acid, 139
— 2,4-dinitrophenylhydrazone, 142
Glycols, acetylenic, 110
Glyoxal, 138, 139
— 2,4-dinitrophenylhydrazone, 142
Glyoxalate, 25
Glyoxylic acid, 138
2′-GMP, 29
3′-GMP, 29
Gradient elution, 93
epi-Griseofulvin, 40
Griseofulvin, racemic, 40
Guaiacol, TLE, 61
Guaran, 27
Gum arabic, 27
Gum tragacanth, 27

Haemoglobin, adult human, 211
Halohydroxy fatty acids, 103
Harmine, 158
n-Heptylaldehyde, 138
Histamine, 21
—, TLE, 64

Histidine, 167
—, TLE, 63, 64
History of development of TLC, 1
Holstine, 156
Horizontal continuous running TLC, 93
Hydrastine, 150
Hydrocarbons, chlorinated, 40
Hydrochinone, 135
Hydrocortisone, 183
Hydrogen bridges, 42
Hydroperoxides, 108
— of ketones, 109
Hydroquinone, TLE, 59, 60
Hydroxy acids, 102
p-Hydroxybenzoate, 109
Hydroxybenzoic acids, TLE, 59
m-Hydroxybenzoic acid, TLE, 59,60
p-Hydroxybenzoic acid, 59, 218
— —, TLE, 59, 60
2-Hydroxybenzophenone, 31
— derivatives, 31
β-Hydroxybutyrate, 25
Hydroxy compounds, 90
Hydroxy fatty acids, 105
α-Hydroxy fatty acids, 103
— — esters, 103
Hydroxyl apatite, 92
9-Hydroxy-12-octadecanoic acids, 103
12-Hydroxy-9-octadecanoic acids, 103
9-Hydroxy-*trans*-10-*cis*-12-octadecadienoic acids, 102
9-Hydroxy-*trans*-10-*trans*-12-octadecadienoic acids, 102
6α-Hydroxy-17β-oestradiol, 84
16α-Hydroxy-oestrone, 84
16β-Hydroxy-oestrone, 84
12-Hydroxyoleic acid, 103
21-Hydroxy-Δ^4-pregnene-3,11,20-trione, 183
3β-Hydroxypregn-5-en-20-one, 78
17α-Hydroxyprogesterone, 183
Hydroxyproline, 167, 171
Hypoxanthine, 28

Impregnation of layers, 91
Indantrene blue RS, 213
Indicators, 40
Indigo carmine, 213
Indophenol, 119
Ink dyes, 23
Inorganic chromatoplates, 80
— ions, 87, 221
— —, TLE, 65
— phosphor layers, 82
Inosine, 28
Insecticides, 110
Iodate–periodate mixtures, TLE, 65
Iodate, quantitative analysis, 66
Iodates, TLE, 65
Iodine vapour, 82, 95
Ion-exchange chromatography, 92
Isoamylamine, 21
dl-Isocitrate, 24, 25
Isoferulic acid, TLE, 61
Isolation of separated substances, 5

Isoleucine, 167
Isomenthol, 53
cis–trans-Isomers, 102

Jasmine extract, 46
Johannisbrotkernmehl, 27

Ketodienes, 109
α-Ketoglutarate, 24, 25
Kieselgel, 14
Kieselguhr, 14
Kojic acid, TLE, 59, 61

Lactate, 24, 25
Lanosterol, 202
Lard, 91
LDH isozymes, 92
Lead, 222
Lecithins, 108
Leucine, 167
9,12-Linoleate, isomers, 102
9,12,15-Linolenate, isomers, 102
2-Linoleodistearin, 107
Lipids, 87, 90, 93, 94, 99
—, extracts, 92
—, highly polar, 98
—, nonpolar, 92
—, radioactive-labelled, 95, 97, 99
—, separation into classes, 99
Lipophilic dyes, 78
— substances, 82
Lobelia alkaloids, 21, 40
Loose powder, TLC, 93
Lysine, 167
—, TLE, 63, 64
Lysophosphatidylcholine, 175
—, hydrolysis products, 107

Magnesium, 149, 220
— carbonate, 92
— oxide, 92
Malate, 24, 25
Malonate, 25
Mannose, 25–27
D-Mannose, 26
Mass spectrometry and thin-layer chromatography, 69
Menthol, 53
Mercuric acetate addition products of unsaturated compounds, 94
Mercury, 222
Mescaline, TLE, 64
Mesoxalic acid, diethyl ester, 138
Metal cations, 20, 23, 24
— —, alkaline earth group, 23, 24
— —, ammonium sulphide group, 23, 24
— —, hydrogen sulphide group, 23, 24
— dithizonates, 221, 222
Methionine, 167
—, TLE, 63
p-Methoxyazobenzene, 38
Methylamine, 21
—, TLE, 63, 64

Methyl brassicate, 102
— elaidate, 102
— erucate, 102
— gentisate, TLE, 59, 60
— p-hydroxybenzoate, 218
— linoleate, 101
— linolenate, 101
— octadecanoate isomers, 102
— oleate, 100–102
— orange, TLE, 65
— palmitate, 100
— red, TLE, 65
— stearate, 101, 102
3-O-Methylsphingosine, 108
Microchromatoplates, 92
Micro-circular method, 75
Microreactions, 82
Microslides, 93
Microsublimation, 82
Mixed solvent systems, 93
Mixotropic series, 75
— — of solvents, 76
Modification of mixtures to improve separation, 94
Molecular structure and chromatographic behaviour, 51
Molybdate, impregnation of layers, 90
Monoglycerides, 97, 104
—, isomers, 104
Monohydroxy acids, 102
Monoolein, 100
Monoterpenes, 40
Morin, 37, 69, 82
Morphine, 127, 150
Multiple development, 106
Multiple thin-layer chromatography, 132
Myristicin, 110

Naphthalene, 133
α-Naphthol, TLE, 62
β-Naphthol, TLE, 62
Naphthols, 94
—, TLE, 62
α-Naphthol-5-sulphonic acid, TLE, 62
Narceine, 150
Narcotine, 150
Neocarmine W, TLE, 65
Neoisomenthol, 53
Neomenthol, 53
Neomycin sulphates, 92
Nickel, 220
Nicotinamide, 109
Nitroanilines, 40
o,m,p-Nitrobenzaldoximes, 160
4-Nitrobenzene-(1-azo-4)-naphthol, TLE, 62
4-Nitro-2′-methyl-4′-diethanol aminoazobenzene, 79
Nitrophenols, 40, 43
Noradrenaline, TLE, 64
Nordihydroguaiaretic acid, 109
Norleucine, 167
Norvaline, 167
Nucleic acid derivatives, 20, 28, 29, 31
Nucleobases, 28

Nucleoside, 28
Nucleotide, 28, 29
Nurugum, 27

Octadecadienoate, isomers, 101
Octadecanoate, isomers, 102
Octadecene-9, 100
n-Octylaldehyde, 138
Odorous substances, 45
Oenanthol 2,4-dinitrophenylhydrazone, 142
Oestradiol-17β, 78, 79
Oestriol, 84
16-epi-Oestriol, 84
Oestrogens, 83
—, polar, 83
Oestrone, 78
Oleic acid, 93, 100, 103
1-Oleodistearin, 107
2-Oleodistearin, 107
Oleyl alcohol, 100
Oleylaldehyde, 100
Oleyl oleate, 100
Olive oil, 91
Ololiuqui, 40
Opium alkaloids, 21, 40
Optically active silica gels, 9
Oracetorange 2 R, CIBA, 79
Oracetred 2 G, CIBA, 79
Orange GGN, 213
— S, 213
Organic peroxides, 109
Ornithine, 167
Osazones, 139
Overflow technique, 35
Oxalacetate, 25
Oximes, isomers, 160
6-Oxo-oestradiol, 84
16-Oxo-oestradiol, 84
Δ^4-3-Oxosteroids, 80, 82
N-Oxystrychnine, 156

Palmitic acid, 93
Palmitoleic acid, 103
Papaverine, 127, 150
Patent blue V, 213
Pectin, 27
Peptides, 91, 208
Periodates, TLE, 65
Perlon, 92
Pesticides, chlorinated, 110
Petroselenic acid, 110
Phenanthrene, 133
Phenanthridine, 135
Phenolic carboxylic acids, 90
— compounds, 20
Phenolphthalein, TLE, 65
Phenols, TLE, 58, 60
Phenylalanine, 167
—, TLE, 63
Phloroglucinol, TLE, 60
Phosphatides, 96, 107, 108, 110
Phosphatidic acid, 175
Phosphatidyl choline, 175

— —, hydrolysis products, 107
— ethanolamine, 175
— —, hydrolysis products, 107
— inositol, 175
— —, hydrolysis products, 107
— serine, 108, 175
— —, hydrolysis products, 107
Phospholipids, 91, 96, 107, 108, 174
Phytosphingosine, 108
C28-Phytosterols, 202
Piperidine, TLE, 64
Plasticizers, 110
Polyacrylonitrile, 10, 92
Polyamide, 10
— powder, 92
Polyethyleneimine, nondialysable, impregnation of layers, 92
Polyethylene powder, 91
Polyhydroxyphenols, 27, 28
Polyphenyls, 40
Ponceau 4 R, 213
Ponceau 6 R, 213
Potassium, 220
Prednisolone, 183
Prednisone, 183
Pregnanediol, 84
5β-Pregnane-3α,20α-diol, 78
Preparative technique, 38
— TLC, 97
Pro-anthocyanidin L$_1$, 27
Progesterone, 78
Proline, 167, 173
—, TLE, 64
Propionaldehyde, 138, 139
— 2,4-dinitrophenylhydrazone, 142
n-Propylamine, 21
Propylenediamine, TLE, 64
Propyl gallate, 109
— p-hydroxybenzoate, 218
Proteins, 91, 205, 209
Protocatechuic acid, TLE, 59, 60
Pseudostrychnine, 156
Publications on TLC, number of, 13
Purines, 82
—, detection method, 82
Pyrene, 133
Pyridoxamine, 211
Pyrocatechol, TLE, 59, 60
Pyrogallol, TLE, 59, 60
Pyruvate, 24, 25
Pyruvic acid 2,4-dinitrophenylhydrazone, 142

Quantitative analysis of macro amounts of periodate and iodate, 66
— determination, 220
— evaluation of TLC chromatograms, 11
— TLC, 96
Quinidine, 150
Quinine, 127, 150
— sulphate, 82
Quinoline yellow, 213
Quinone, 109, 135
—, TLE, 59, 60

Rauwolfia alkaloids, 21
Recording chromatograms, 96
Recovery of substances from Sephadex layer, 208
Relation between molecular weight and chromatographic mobility on Sephadex G 100, 211
Relationship between R_S and number of carbon atoms, 140
— — spot area and concentration, 97
Relative merits of paper and thin-layer chromatography of steroids, 86
Resorcinol, TLE, 60
—, dimethyl ether, TLE, 61
α-Resorcylic acid, 60
— —, TLE, 60
Retuline, 156
Reversed-phase chromatography, 91, 93, 100
— technique, 79
— TLC, 106, 195
R_F values, reproducibility, 94
Rhamnose, 25–27
Rhodamine B, TLE, 65
Ribose, 25–27
D-Ribose, 26

Salicylic acid, 218
— —, TLE, 60
— aldehyde, TLE, 60
Salt formation, 42
Scarlet GN, 213
S-chamber, 7
Selachyl alcohol, 100
— diolein, 100
Separation chambers, 6
— according to degree of unsaturation, 101
Sephadex, 10
—, adsorbent in TLC, 205
Serine, 167
Sesquiterpenes, 40
Silica gel, 10
— — with chelate-forming anions, 90
— — G, 9
— — H, 9
Silver nitrate, impregnation of layers, 90
β-Sitosterol, 198, 202
Skytanthine, 155
α-Skytanthine, 156
β-Skytanthine, 156
δ-Skytanthine, 156
Sodium, 220
— alginate, 27
Solvents, for steroids, TLC, 75
Sorbents for TLC, 10
—, inorganic, for TLC, 10
—, organic, for TLC, 10
Sorbic acid, 218
Sorption media, 7
Sphingolipids, 107
Sphingomyelin, 108, 175
—, hydrolysis products, 107
erythro-Sphingosine, 108
threo-Sphingosine, 108
Sphingosine bases, 108

— derivatives, 108
Spray gun, 88
Starch, 117
Stereoisomers, separation, 155
Steroids, 40, 75
—, carbonisation reaction, 82
—, detecting reagents, 81
—, fluorescence reactions, 80
—, hydrophilic, 75
—, lipophilic, 75
—, neutral, 78
—, phenolic, 78, 82
—, quantitative evaluation, 80
—, weakly basic, 78
Sterol acetates, 197, 198
Sterols, 110
Stigmasterol, 198, 202
Stilboestrol, 195
Structure and R_F values, relationship between, 134
Strychnine, 150, 156
Strychnos alkaloids, 156
Substituents, influence on the R_F value, 79
Succinate, 24, 25
Sudan red, 38, 119
Sudan yellow, 38
Sugars, 10, 20, 25–27, 31, 40, 92
Sugar acids, 24
Sulphosalicylic acid, 82
Surface chromatography, 8
Syringic acid, TLE, 57, 58, 61
— aldehyde, TLE, 61

Tartrazine, 213
Taurine, 167
Terpenes, 87
Terpenoid hydrocarbons, 90
Testosterone, 78, 79
Tetrahydroacenaphthene, 136
Tetrahydrocortisol, 78
3α,11β,17α,21-Tetrahydroxypregnan-20-one, 183
TLC on microslides, 89
— in test tubes, 89
TLE, 55
Thebaine, 150
Thin-layer applicator, 5
Threonine, 167
Thymolphthalein, TLE, 65
Tocopherol, 40, 97, 109, 110
—, mixtures, 109
α-Tocopherol acetate, 40
p-Tolualdoximes, 160
Triamcinolone, 183
Tricarbonic acid cycle substrates, 20, 24, 25
Triethanolamine, TLE, 63
Triglycerides, 97, 98, 104
—, brominated, 91
—, castor oil, 105, 198
—, cocoa butter, 106
—, corn oil, 105
—, critical pairs, 105, 106
—, fractionation according to degree of unsaturation, 106

Triglycerides *(contd.)*, groundnut oil, 105
—, human blood serum, 105
—, lard, 105, 106
—, linseed oil,
—, Malayan palm oil, 106
—, mono-acid unsaturated, 105
—, olive oil, 105
—, sesame oil, 105
—, shea butter, 106
—, soya bean oil, 105
—, unsaturated, 93
$3\beta,11\beta,21$-Trihydroxyallopregnan-20-one, 183
$3\alpha,17\alpha,21$-Trihydroxypregnane-11,20-dione, 183
$11\beta,17\alpha,21$-Trihydroxypregnane-3,20-dione, 183
$3\alpha,17\alpha,21$-Trihydroxypregnan-20-one, 183
Trilinolenin, 100
Triolein, 100, 107
Tristearin, 107
Triterpenic compounds, 40
Triterpenoid acids, 102
Tryptophan, 167, 172
—, TLE, 63
Tungstate, impregnation of layers, 90
Types of layer, 89
Tyramine, 21
Tyrosine, 167

Ubiquinones, 96, 110
Umbelliferone, TLE, 61
3'-UMP, 29
Unsaturated acids, 101
— —, brominated, 95
— —, fluorescein–bromine test for, 82
— —, hydrogenated, 95
Uracil, 28
Uridine, 28
Uronic acid, 20, 24, 26

Vacuum cleaner, 6, 97
n-Valeraldehyde 2,4-dinitrophenylhydrazone, 142
Valine, 167
Van der Waals forces, 42
Vanillic acid, TLE, 61
Vanillin, TLE, 61
iso-Vanillin, TLE, 61
o-Vanillin, TLE, 61
Veratric acid, TLE, 61
Vinca minor alkaloids, 40
Vitamin A, 40, 110
— acetate, 40
— esters, 109
— isomers, 109
Vitamin A_2 isomers, 109
Vitamin B_1, 109
Vitamin B_2, 109
Vitamin B_6, 109
Vitamin C, 109
Vitamin D_2, 40, 109, 110
Vitamin E, 109
Vitamin K, 110
Vitamin K_1, 109
Vitamin K_2, 109
Vitamin K_3, 109
Vitamins, 90
—, fat-soluble, 109
Voacanga africana alkaloids, 40
Volatile compounds, 98
Vomicine, 156

Wedge strip technique, 93
Withaurea somnifera alkaloids, 40

Xanthophyll degradation products, 109
Xylose, 25-27
D-Xylose, 26

Zinc, 222

PRINTED IN THE NETHERLANDS